T0143431

Manufacturing of Quality Oral Drug Products

Manufacturing of Quality Oral Drug Products

Processing and Safe Handling of Active Pharmaceutical Ingredients (API)

Sam A. Hout

CRC Press is an imprint of the
Taylor & Francis Group, an informa business

First edition published 2022
by CRC Press
6000 Broken Sound Parkway NW, Suite 300, Boca Raton, FL 33487-2742

and by CRC Press
4 Park Square, Milton Park, Abingdon, Oxon, OX14 4RN

CRC Press is an imprint of Taylor & Francis Group, LLC

ISBN: 978-1-032-12473-5 (hbk)
ISBN: 978-1-032-12474-2 (pbk)
ISBN: 978-1-003-22471-6 (ebk)

DOI: 10.1201/9781003224716

Typeset in Times
by codeMantra

Contents

Preface

I conceptualized writing this book on drug manufacturing quality covering oral solid dosage (OSD), coatings technologies, and capsule filling (liquids and powders), with an emphasis on API safe handling and analytical method applications after spending over 30 years in drug manufacturing engineering, including aseptic liquid and lyophilized parenteral drug vial fill/finish, cartridge filling for injector applications, sterile prefilled syringes, and ophthalmic bottle filling and copping. Quality by Design (QbD) dictates process preparations and compounding of drug formulations are key steps in transferring bulk product batches from beginning to end. These processes all require diligent care and technical know-how in transferring across process steps while maintaining clean and safe operations. Personnel flow and materials flow separations and consideration for movement have rigorous requirements to ensure against cross contamination and making a safe product. There are many guidelines that govern all these processes defining specific requirements. I wanted to put it all together in one accessible simplified experience-based reference to reinforce knowledge of compliance regulations and processes, and provide an easy guide to follow. This book is important in providing a step-by-step understanding of what is required to engineer and manufacture drug products.

Process flow, product risk-based quality approach, and the significance of process validation are emphasized. Critical step-by-step technologies and systems are described as part of the integrated facilities design. API safe handling and analytical methods techniques to facilitate process, product testing, and reliability in batch-to-batch consistency and repeatable production in a compliant manner are discussed. Handling both API transfers and high potency drugs are key developments in this book. The systematic management of risk to ensure efficacious and safe pharmaceutical is explained in terms as it relates to process validation. The importance of change control and analytical testing of raw material quality and in-process testing for both chemistry and microbiology evaluations is systemic in drug manufacture.

The purpose of this book is to help my colleagues in the drug manufacturing industries have a simple reference book on important issues that we deal with daily and always looking for ways to troubleshoot and solve problems that face us in manufacturing drug substances (API), excipients, and products.

To develop this book, I had to learn more and more about the subject matter that I am describing and covering, and I had to investigate insights and real examples about situations that I had difficult and complex formulations that required specific innovative ways to process while ensuring compliance. I believe that through learning, training, and asking questions about alternatives, I was able to grow in how to convey this message in a more concise and focused fashion.

The feedback from the reviewers opened my thinking to how others are perceiving the materials and required to add value in terms of case studies and real examples from experience.

Besides writing from my personal experience in pharmaceutical industry, I had to research and follow guidelines from FDA, MHRA, EMA, ISPE, and ICH, among other references, that have been reviewed and cited as part of this publication.

I had many communications with colleagues to see how ideas can be explained and described in the desired methods. I have been thinking about how to aid on technical subjects that are pertinent to this field for over 20 years, and I started putting together notes that would help my colleagues in the industry for over 10 years.

The table of contents in this book flows as intended for the reader to explore subject matter in a continuous fashion or refer to a specific section as needed. The main compliance message in this book is that FDA inspections focus specific concentration on two areas: cGMP compliance and PAI for NDA. Documentation and procedures to cleaning validation, pest control trending, overall facilities adherence to SOP and organizational structure, including appropriate protocols and reports covering all aspect of a product submission for pre-approval, namely, exhibit batches data, stability batches data, microbiology, and chemistry data – all are presented in an orderly manner with full documentation including all CAPAs and deviations.

Acknowledgments

This book is written with the support of my family, my caring loving wife Mona, my empathetic daughter Samantha, and my wise son Owen. Modestly, many colleagues and friends in Lebanon, England, and the US have played a transformational role in adding to my knowledge and experience to add value in what I learned and how to share in the service of others. My colleagues at J&J, Teva, and Siegfried contributed to my project management, engineering and execution capabilities, manufacturing science, and specific aspects of pharmaceutical processes and compliance through our many communications over three decades. Special dedication is hereby expressed to my Uncle Dr. Wally A. Barakat and Aunt Shirley Barakat for their guidance and nascent support to our family endeavors and growth in the US. Special thanks are due to my colleagues: Frank Pham for his spiritual devotion, contributions, and support in set up of projects financial cost controls and overall revenue analysis; Steve Hong for his loyalty, detailed editing of Capex project write-ups, and operational execution excellence; and Tuyet Nguyen for his friendship, creative mechanical fixture design, and troubleshooting capabilities. I am also grateful for team building friendships of Dr. Andre Krol and John Slape that emphasized a fun motivational spirit in the face of challenges for more than three decades.

Introduction

Controlled-release drug product solid dosage involves the movement of water through a semipermeable membrane. Rate-controlled delivery reduces side effects, increases efficacy, and decreases dosing frequency, which improves patient condition. Conventional OSD delivers drugs very quickly. Their efficacy is dependent on the amount of food in the stomach and the acidity of the stomach fluids. Controlled-release solid dosage slows drug delivery dissolution rates using various coatings. Controlled-release OSD systems are dependent on the osmolarity of the surrounding body fluids and temperature. These systems are permeable to water, and the release of active ingredients is dependent on the local conditions in the digestive tract due to coatings. The system continues to release its contents even after it has moved into the small intestine. The release process is dependent on the solute concentration within the OSD is greater than that of its surroundings.

Controlled-release system has a solid core containing therapeutic and osmotic agents. This core is coated with a semipermeable rate-controlling membrane that might have one or more laser-drilled orifices. This system mostly has a water-soluble drug and an osmotic agent in a monolithic core and delivers the drug in solution. In addition, multilayers have a separate osmotic characteristic and the drug delivers in solution or suspension. This system can deliver either water-soluble or insoluble drugs. In the moist environment of the GI tract, water enters the system through the semipermeable membrane and dissolves or suspends the drug core. The drug is released through the orifice(s) at a rate that is controlled by the osmotic and chemical properties of OSD structure and the membrane's permeability to water. This system provides a constant rate of drug delivery.

Controlled-release system is dependent on the following variables:

1. ingredients
2. permeability of the membrane coating
3. solubility of the overcoating and a sub-coating
4. size and number of the drilled orifices.

Controlled-release system mechanisms are based on osmosis – the movement of water through a membrane by diffusion. Water moves toward areas of higher solute concentration until the solutions on either side of the membrane have equal concentrations. The osmotic potential is determined by relative concentration of the solutions on either side of the membrane. The greater the difference in concentrations, the greater the osmotic potential.

The strength of osmotic pressure is controlled by:

1. permeability of the membrane to water
2. solubility of the components in core
3. concentration of solutes within the system
4. consistency of the material being discharged from the system
5. the size/number of the discharge orifice(s).

The permeability of the membrane is based on its composition and thickness. The choice and optimization of the membrane material account for a considerable portion of the compound performance. Membrane not only controls the release of the active substance, but also must ensure a constant volume for the system, as well as being resistant to mechanical stress and digestive fluids. Solute levels, which affect the osmotic potential, are adjusted using sodium chloride in the formulation of the OSD. The consistency of the drug component in the system is controlled by the solubility of the drug and the physical and chemical characteristics of the other excipients in the mix.

An elementary osmotic pump system (single layer) and the expanding material that contains the active drug are both expelled from the system into the GI tract. In a multilayer system (bilayer), the osmotic drug layer gels and expands while the osmotic layer expands. The combined force expels the drug-laden gel into the GI tract. Viscosity of each layer is based on the nominal molecular weight (size/length) of the polymer used in its manufacture. If the orifice were to become temporarily blocked, by a particle or undissolved core component, hydrostatic pressure within the drug reservoir would rise until it was enough to clear the blockage.

Excipients can be divided into two general categories:

- Compression characteristics variables – fillers, diluents, binders, adhesives, lubricants, anti-adherents, and glidants
- Biopharmaceutics stability – disintegrant, polymers, resins, colors, flavors, buffers, and absorbers

Controlled-release manufacturing process involves the following steps:

1. Milling and charging, where components are received, milled into fine powder, and sifted through screens into the initial production vessels
2. Granulation, the mixing of powdered components with moisturizing and binding agents and then blending with dry lubricants
3. Compression of the granulated mixture into solid cores
4. Sub-coating the cores, if a time delay is needed, by applying a water-soluble coating
5. Membrane coating by applying the organic solvent-based semipermeable membrane
6. Laser drilling of the drug delivery orifice in the membrane
7. Drying the drilled cores to remove solvents used in membrane coating
8. Drug and/or color overcoating the drilled cores
9. Printing drug-specific information on the tablets
10. Sorting tablets to remove under- and over-sized tablets
11. Packing the tablets for shipment to clients.

Controlled extended-release system is used for the management of hypertension and angina drug products such as calcium channel blockers (e.g., Nifedipine). Drug delivery delay is a result of an aqueous sub-coating applied to the core before membrane coating. After delay, the drug is delivered using a membrane coating system over a 24-hour period. Other examples of extended-release OSD applications are for the treatment of hypertension using an antihypertensive that dilates blood vessels utilizing a bilayer coating system.

Authors

Sam A. Hout, PhD, MBA, is a Chartered Chemical Engineer, certified in business management by the American Production and Inventory Control Society (APICS), and a member of the International Society of Pharmaceutical Engineers (ISPE). Dr. Hout is VP, Technical Operations at KC Pharmaceuticals. He also held the position of Senior Director of Engineering, project management, and technology process transfers at Siegfried Pharmaceuticals. Previously, he was Senior Manager of Engineering at TEVA Pharmaceuticals and Director of Operations at the HPLC company Phenomenex. Earlier background includes Johnson & Johnson and US Federal Research.

1 Milling and Charging

Raw materials are received by the warehouse and placed in quarantine until released. To be released an item must:

1. be from an approved vendor
2. meet Quality Standards (AQS), which might require examination/testing by materials lab
3. be entered in manufacturing planning and control database. Powdered materials used in granulation are dispensed from the warehouse's pharmacy and staged near milling and charging area.

The first step is to mill materials specified in the Master Formula (MF), usually only sodium chloride and ferric oxide require milling. However, other components can be milled if required. In the milling process, materials are ground by, e.g., Quadro Comil (conical mill) equipment, or possibly by a fluid air-mill, and passed through a sieving screen to break up lumps and to ensure a reduced and uniform particle size for proper blending and granulation. Materials that require milling tend to clump, so they must be protected from moisture by sealing them in poly bags and keeping them away from moisture. They must be charged/mixed with the rest of the powdered materials within a specified time, typically within 72 hours of milling.

Charging involves loading pre-weighed materials received from the pharmacy into tote bins or directly into blender, e.g., Patterson-Kelley (PK). All components are charged using sifting screens, e.g., Porta-Sifter with a screen

1. break up clumps
2. ensure a uniform consistency
3. identify and remove possible foreign material.

Tote charging is accomplished in the designated milling-and-charging room. A lift (e.g., Meto) is used to empty barrels into the Porta-Sifter, which is on an elevated platform above the totes. If the blender is manually charged directly from barrels, then the portable Porta-Sifter screens the material directly to the blender.

MILLING

Materials, such as sodium chloride and ferric oxide, require milling per the MF. They are milled using the Quadro Comil. Quadro Comil is fitted with a 20-mesh screen per MF (mesh number stands for the number of openings per inch in the screen). Powders are passed through the mill and collected in poly bags.

Based on common practice, the pharmacy dispenses slightly more materials than required, since some loss is expected during milling. The operators divide and weigh

DOI: 10.1201/9781003224716-1

the milled powders based on MF-specified quantities for each bowl (tote) being prepared. A bowl represents the quantity that will fit into the agglomeration equipment (e.g., Glatt fluidized bed).

CHARGING

Materials that do not require milling are received from the pharmacy pre-weighed and divided into quantities required for a bowl. After charging to totes, the net weight of the totes is determined to ensure that the bowl weight falls within the validated loads and that all ingredients were charged.

If required, the lot is split into bowls, usually due to the size/weight limitations of the Glatt fluid bed granulator (FBG). Materials for each bowl are placed in totes in the MF-specified order.

Meto-Lift is designed to lift and invert barrels over the Porta-Sifter. It has a funnel-like top that clamps over the open barrel. The flow of material is controlled by a gate valve in the funnel opening. After the barrel has been inverted and attached to the Porta-Sifter using a rubber collar, the valve is opened allowing the granulation to flow from the barrel.

The Porta-Sifter is on an elevated platform that allows totes to be rolled under it. It is fitted with a six-mesh screen (six openings per inch). The granulation flows through the Porta-Sifter into the tote. This screening helps break up any clumps that may have formed in the granulation and helps to ensure that foreign materials are not included in the granulation.

If blender cannot be charged from totes, it is possible to charge it directly from barrels through a portable Porta-Sifter. In this case, attention must be paid to the oxygen level as the powders can be explosive. Anytime the blender is opened while processing materials, it must be purged with nitrogen. When loading from barrels, the oxygen level should be checked after each barrel, and purging should be done whenever the oxygen level reaches or exceeds 8%.

Milled sodium chloride and ferric oxide (if required) must be added to totes/blender within 72 hours (3 days) of milling. Typically, raw materials may be stored in totes for a maximum of 120 hours (5 days) before the start of granulation. Net weight of the tote must be determined and compared to the specified values in MF. Work in progress (WIP) totes must be labeled with in-process labels showing weights, lot numbers, and dates. Totes are moved to staging area for granulation.

If components are received, already granulated, in barrels from client/supplier, it is put into totes using the Meto-Lift and Porta-Sifter, has the lubricant added/blended, and then goes directly to compression process. Totes are labeled with a WIP label showing weights, lot number, and dates. After tote charging of both drug and osmotic granulation, totes are taken to blending, then to the staging area for core compression. Critical elements of the process are:

- Failure to mill and screen powders could result in poor distribution of materials throughout granulation. This could have an impact on both core compression and drug dosage levels.

- Even dispersal of sodium chloride is critical to the osmotic properties of granulation since it is used to regulate the solute levels of granulation.
- Even distribution of ferric oxide is important in getting a consistent coloring in the core OSD.

Totes should be thoroughly checked before charging to ensure that they are clean, dry, and contain no foreign materials. Dump valve seal check by opening and closing the valve should reveal condition of the seal. Operators should always check the screen after charging each barrel for the presence of foreign material. The barrels, poly bags, ties, and seals should be kept away from the charging area. The use of razor blades in dispensing can create problems as they can easily fall into the materials or shred plastic bags.

An easy way to cross-check what is in the tote is to take the total weight of all components given on the MF and compare it to the net weight of the tote. If the weights differ, it could indicate that either not all or extra material were charged, or that there is a problem with pre-weighed materials from the pharmacy.

2 Granulation

Granulation is a process of agglomeration, whereby smaller particles are brought together into an aggregate with increased void space in the larger aggregate free-flowing state to allow for breakdown into dissolution when hydrated. The process of granulating is one of the most versatile and important precompression process steps in drug product solid dosage manufacturing. By granulating, powders can be mixed, and cling together to increase free-flowing characteristic, instantize dissolution, and become nearly dust free.

Granulating involves the blending of powdered components, the introduction of a binding agent, drying the resulting mixture, and milling to obtain a uniform granule size. Dry powders are moistened with a solvent during granulation. Some of the powdered material dissolves at least partially. A process aid binder is added to the dry mix to improve adhesion when in solution. Pendular bridges are formed between the particles during granulation. The open compressible structure allows the granule to hold moisture internally yet still be free flowing. Granules can be formed as a dry mix, pendular, funicular, capillary, kneaded capillary, and coated. The tensile strength increases about three times from pendular to the capillary state.

Dry granulating applications require added pressure and/or heat to form slugs, which are then milled to a uniform size. Wet granulating of dry mixtures, whereby they are mixed with a solvent to dough consistency and then dried and milled to achieve a uniform granule. Many of these processes involve moving the product between various vessels as it moves from blending, to agglomerating, to drying, and to milling. This approach is inefficient in terms of process times and labor costs, and it exposes the product to potential contamination.

There are two different methods to prepare powdered materials by using fluid-bed granulator or a twin-shell liquid-solid blender. The first three steps – mixing, agglomerating, and drying – in both processes are accomplished in a single vessel.

A fluid-bed granulator (e.g., Glatt) suspends the powder particles in a stream of air, a fluidized bed, while atomized binder solution is sprayed. It uses heated air to dry the granulation throughout the process. A blender (e.g., PK) mixes the powders with an organic solvent, ethanol using a tumbling (rotating) motion. It then uses heat and a vacuum to remove ethanol from the granulation at the end of the process. In both methods, the granulation is milled for size uniformity at the end.

Granulation allows for instantized particle size and weight for increased dissolution and reduction of dust levels, which improves powder explosion risk. This allows safer, cleaner, and easier handling with less loss of product due to airborne or electrostatic bonding to manufacturing surfaces.

Agglomeration (granulation) fuses particles of different size, shape, and density permits easier mechanical handling without a substantial loss of mix quality due to segregation. The fusing of particles has little or no effect on the chemical structure of the components; it helps to ensure a consistent blend of all components.

DOI: 10.1201/9781003224716-2

Agglomerates facilitate mixing by improving homogeneity of low dosage level products by evenly mixing and adhering the active drug to fillers and diluents. In addition, it improves dosage uniformity by improving the uniformity of components that reach the compression machine by decreasing segregation, which is due to differences in density and size.

PK BLENDER MANUFACTURING PROCESS

The blender intensifier bar is used to enhance the spray pattern homogeneity during the spraying cycle. It rotates at approximately 3,000 rpm. Dispersion blades keep the granulation from sticking to the spray discs. The entire intensifier bar assembly is removed and disassembled for cleaning. The blending process uses an organic solvent, anhydrous ethyl alcohol, and a binding agent, HPMC, which is added to dry ingredients before spraying. The process forms pendular bridged granules. During spraying, ethyl alcohol dissolves HPMC, which allows it to form the bridges between particles through surface tension/capillary action. The bridges then recrystallize during drying. The API drug substance and other excipients are not soluble in ethyl alcohol. At the end of spraying, the mixture has a moist loose granular consistency, which easily compacts much like a snowball. It has a moisture level of 18%–24% by weight depending on the product. The granulation is then dried by heating the jacketed vessel to a temperature between 70°F and 100°F per MF and creating a vacuum within the vessel. The vacuum applied is normally 100 mm Hg (standard ambient atmospheric pressure is 760 mm Hg). Drying times can take as long as 24 hours. The vacuum pumps and alcohol condenser are usually located outside the blender building.

Fluidized-bed granules are relatively uniform in size, even before milling. Blender granulation can be lumpy. As a result, milling to a uniform size granule is critical. The blender vessel is inverted with the charge/discharge hatch down, and a fluid air mill is attached. As the granulation is milled, it is moved to a tote with a spiral feeder. The portable fluid air mill is continuously purged with nitrogen and is supplied with chilled water. The maximum allowable temperature is 100°F. After milling, the granulation is pre-blended on the tote tumbler and then the lubricant, magnesium stearate, is added to the tote and finely blended.

CHARGING

Materials are received, and the anhydrous ethyl alcohol is dispensed into pressure vessel through a 3-μm filter. Air is then used to pressurize the vessel and push the alcohol through the lines to the blender spray discs. The blender can be charged directly from drums by using a portable Porta-Sifter. Powders are sifted directly into the blender in the order specified in MF. While charging the blender, attention must be paid to the oxygen level as the powders can be explosive. Anytime the blender is opened during processing, it must be purged with nitrogen. When charging from barrels, the oxygen level should be checked after each barrel; if the oxygen level reaches or exceeds 8%, blender must be purged.

To charge from totes, the blender is rotated so the neck is aligned with the opening in the ceiling. It is purged with nitrogen and the charging collar is attached to the top of PK blender. The tote is positioned in the charge station over the PK and the tote is grounded. The inflatable seal on top of the collar is inflated to form a seal between the PK collar and the tote. The tote is then purged with nitrogen and the valve on the bottom of the tote is opened. To ensure that all the powders fall into the PK, the operators can use a slight vacuum and/or they can tap on the sides of the tote using rubber mallets. The same procedures are repeated for the second tote. After charging, components are blended by rotating the PK at 6 rpm for 30 minutes with intensifier bar off.

Granulating

The intensifier bar is turned on and a specified quantity of ethyl alcohol is sprayed while blending for a specified time. Most products call for the spraying to occur in several increments, normally with the bulk of the alcohol being sprayed during the first one. Spraying is complete when the wet sample shows a moisture level within range for that product (normally 18%–24%) per MF. When the required moisture level is reached, the hot water circulation pumps are started, and the MF-specified temperature is set. Normally the filter in the vacuum line is changed. However, in some cases, the pressure differential across the filter is monitored, and the filter is changed only when it reaches a given limit. A new circle chart is installed to monitor the vacuum, and the vacuum pumps are started. The vacuum is taken down slowly to 100 mm Hg.

With the intensifier bar off, the first 2 hours of drying involves cycling the blender rotation between 1.5 rpm for 5 minutes and 0 rpm for 10 minutes. After the first two hours, the PK is set at 1.5 rpm, and the granulation can dry for 8–10 hours depending on the product. After the second drying cycle, samples are taken (normally from each side of the PK), and the average moisture loss on drying (LOD) is checked against the target. Drying continues until the LOD values are within range. Some products require up to 40 hours to dry. When the desired LOD is reached, the water heater and circulating pumps are turned off.

Milling

The relative humidity of the process room should be maintained below 55%. Blender is positioned with the charge/discharge opening straight down and a portable fluid air mill is connected to it by a cloth boot. The mill and PK are continuously purged with N2, and the mill has chilled water circulating through it to keep it below 100°F. A ten-mesh screen (ten openings per inch) is used on the mill per MF. The walls of the PK should be scraped, and that material is also milled. The outlet of the fluid air mill relates to a cloth boot to the feed hopper of a spiral feeder, screw mechanism to lift the milled granulation up and into a tote.

Final Blending

After milling is complete, the granulation is pre-blended on the tote tumbler at 8 rpm for either 5 or 10 minutes to loosen any clumps before final blending. Magnesium stearate is screened through a 20-mesh screen per MF and added to the top of the tote.

It is dispersed into the granulation with a paddle. The tote is then blended at the MF-specified rpm and time on the tote tumbler.

Post Blending

For osmotic granulation, LOD is checked again to ensure that it is under 1%. Sieve analysis and tap density tests are performed. The SOPs for the various processes specify the sieve screen sizes and specific testing procedures. Accountability check is performed, and totes are moved to compression staging area.

LABORATORY ANALYSIS OF GRANULATION

Samples of granulation are assay tested by commercial lab for content (% active drug), content uniformity particle distribution, and purity.

CRITICAL ELEMENTS OF THE PROCESS

Explosive safety is of critical importance in dealing with the PK blender because of ethanol that is sprayed, and the very fine powders being mixed. Proper grounding, purging the vessel with nitrogen, and monitoring oxygen levels must be a priority. The spraying amount and blend-time intervals used in the process need to be followed closely. Slow dissolution of HPMC and formation of bridged granule is an important step in the process. Spraying the full amount all at once could result in coating instead of forming bridges.

If the valves are opened too quickly at the start of drying, the condenser can flood and would require draining by maintenance personnel. It is possible for the condensed alcohol to be drawn back into the PK blender if the vacuum is not released and/or if the valves are not closed when maintenance tries to drain the condenser.

Drying to the correct moisture level in the blender is just as critical as with the fluid bed. Because of the very long drying times and low moisture content, if the relative humidity is too high (above 55%) the granulation can absorb water from the air and become too moist. This is likely during milling and transfer to tote.

Because the portable fluid air mill is used with the PK, a possibility exists for the wrong size screen to be installed. Multiple screens are used, so always verify that the proper screen is installed per the MF.

The spiral feeder should be operated when product is available to move freely through the tube. If it is left running and there isn't any product in the hopper to push what is in the tube up and out, it could overheat and cause the product to extrude, melt, and harden. Over-blending of magnesium stearate (lubricant) is as significant a problem for blender or fluid-bed granulation.

3 Compression

Following final blending, granulation is compressed into OSD using tablet presses (e.g., Rotapress made by Manesty, Kikusui, or Fette tablet presses). Presses can produce bi-layer and single-layer tablets. Tableting involves the application of a significant compressive mechanical load to a bed of powder resulting in compaction. The physics of compaction involves the compression and consolidation of a two-phase system (particulate solid and gas) into a denser system due to applied forces. Compression is the increase in bulk density as a result of displacement of the gaseous phase by a solid. Consolidation is an increase in mechanical strength resulting from particle-particle interactions.

Compression resulting from the application of an external force decreases the volume, thereby increasing bulk density, by initially causing a closer re-packing of particles and displacing the gases. Increasing the load further will then result in elastic deformation (bending) and eventually plastic deformation (brittle fracture) when the elastic limits are exceeded. However, compression is only the beginning of the process. It is consolidation that really gives the tablet enough strength to withstand the pressures of further manufacturing. During consolidation chemical bonds are formed on the solid surfaces of the particles, a phenomenon called "cold welding." Under the pressures applied to tablets, an additional phenomenon called "fusion bonding" can take place. It is the bonding of particles through the localized melting, due to the heat generated by friction in the particle-particle contacts under the applied force, and the re-solidification when the force is relieved and the particles cool. This consolidation process is influenced by:

1. chemical nature of the particle surfaces
2. amount of available clean surface
3. presence of surface contaminants
4. inter-surface distances (amount of gases present)
5. magnitude of the applied force
6. amount of time the force is acting on the particles

There are three essential characteristics of a granulated product in preparation for compression:

1. free flowing
2. binding properties
3. does not stick to the punches and dies

The size and uniformity of the granulation affect the solid-to-gas ratio of the precompression blend. The excipients, the moisture level, and how well they are blended will affect the surface chemistry and bonding characteristics during consolidation. In addition, the structure of the granule affects the amount of clean surface exposed when fractured, the amount of trapped moisture, and its strength. The pendular structure requires

DOI: 10.1201/9781003224716-3

less force to break apart than a capillary or coated granule. The more lattice-like the granule, the easier it is to fracture. Based on this, over-blending the lubricant could interfere with the consolidation process by lubricating surfaces that should be bonding. Not only are the adhesion characteristics of the lubricants different, but they also reduce the particle-particle friction that produces the heat needed for fusion bonding. Chemical characteristics of formulation, flow characteristics, adhesion, and lubricity dictate the level of press performance in producing acceptable OSD.

Force and time are controlled during tablet compaction. The amount of force applied by the press during compression and consolidation determine

1. tablet's porosity (ratio of solid to gas)
2. amount of brittle fracturing that exposes clean bonding surfaces
3. amount of heat generated by particle-particle friction.

The length of time that the pressure is applied is called the dwell time. Dwell time is controlled by the speed of the press, the diameter of the compression roll, and flat head surface of the punches.

Manesty press tableting can be broken down into seven distinct steps as the turret and tooling move around the press. In the case of single-layer tablets, a tablet is produced in each 180° rotation. For a bi-layer tablet, it takes an entire rotation with each layer being filled and pressed in 180°.

1. Creating cavity in the die:

   Following ejection of the previous tablet, the lower punch is pulled down by the fill cam. This action takes place under the feed frame and helps creating a negative pressure that pulls granulation into the die's cavity. The depth of fill is determined by the size of the fill cam fitted to the press, either full, medium, or shallow. The fill is the first step in determining the tablet's weight.
2. Filling the cavity:

   Granulation fills the die by suction from Step 1, gravity, and the force-feeding action of the rotating paddles within the feed frame.
3. Adjusting the volume of granulation in the die cavity (dosing):

   This is accomplished by weight adjustment, cam, and ramp. The ramp guides the lower punch head as it moves from the fill cam to weight adjustment assembly. A flight control cam prevents the heads of the lower punches from flying upward as they pass over the weight adjustment cam at high speeds (no bouncing allowed). The weight adjustment assembly is used to set the volume of granulation in the die (this establishes the tablet's exact weight). The initial over-fill is necessary to ensure a consistent volume of granulation. However, the amount of over-fill should be kept to a minimum to avoid moving the granulation more than necessary. A 2–4 mm over-fill is usually enough. A spring-loaded scraper removes material that escapes from the feed frame and pushes it into the re-circulation channel in the center of the die table. This is where the correct choice of fill cam is vital. Too much granulation pushed back into the feed frame will cause loss of powder as it is pushed out, since the recirculation channel may not be large enough to carry all the granulation. This can result in a choked feed frame or the loss of granulation and lower yields (less tablets per lot). Too shallow a fill cam and the dies may be only partially filled with low weights and/or weight variation may occur.

4. Reducing the height of the fill to below the die bore chamfer:

   Once a precise volume of powder has been achieved, it must be contained until it can be compacted. This is done by lowering the lower punch with the pull-down cam. This eliminates the loss of granulation from puffing as the upper punch enters the die bore. In addition, use of a "tail over die," a phenolic plate, to keep the filled die covered until just before the upper punch enters the die. This keeps the granulation in the die until the upper punch enters the die, which helps to ensure weight consistency because granulation can be pulled or blown out of the die by air turbulence behind the scraper and/or feed frame. The lower punches are also equipped with an anti-flight system, which controls erratic punch travel due to direction changes caused by the lower cam track while granulation is in the die.

5. Precompression (expelling of air from between the granules within the die):

   This facilitates compaction and reduces tableting problems due to the elastic rebound effect of trapped air, the granulation is de-aerated or predensified. This is accomplished by a set of precompression rolls set to apply a minimum force to the granulation enough to expel most of the trapped air from between granules, normally around 200 Ibs, but always less than 2,000 Ibs. For bi-layer tablets, it is used to compress drug layer to make room for the osmotic granulation. A very soft friable tablet is formed by this process. It is used only if laminating occurs in the formation of single-layer tablets.

6. Compacting the granulation into a strong hard tablet:

   The main compression rolls apply the final consolidation force to the upper and lower punches. This force gives the tablet its strength/hardness by fracturing and bonding of particles. It determines the tablet's hardness and thickness. The consolidation is dependent on the dwell time. Dwell time is the period that the maximum consolidation force is held constant. It is the period that the head flat is under the main compression roll. Dwell time is an expression of head flat diameter divided by the punches' linear velocity.

7. Ejection of the finished tablet from the die:

   This is accomplished by the ejection cam. When this cam is correctly set, the punch tip should be less than 0.002″ above the die table at the point of ejection. If set too low, the tablet may stick in the die and end up broken, or if set too high, it could catch on the take-off blade that removes the tablet from the table. Depending on how much the tablet binds in the die, the force on the ejection cam can be very high. To prevent damage, it must be kept in good condition and well lubricated. Magnesium stearate helps lubricate the die walls facilitating tablet ejection. Over time, the die bores can take on a barrel shape, which greatly increases the ejection force required to remove a tablet. The increased force can result in excessive vibration and wear on the cams and punches. Tablets can also chip or break while being ejected. If the barreling is not excessive, it may be possible to raise the lower punches and compress the tablet higher in the die or flip the die over.

These steps summarize the tableting process for a single-layer tablet. However, if set up is to produce a bi-layer tablet, then a full cycle (360°) around the press is required. On the first 180°, the die is filled with the active (drug) granulation and goes through precompression. The main compression rolls are raised, and the ejection cam is removed. This produces a soft friable tablet composed of just the active granulation. During the final 180°, the osmotic granulation is filled on top of the precompressed drug layer, and the tablet is compacted by the main compression roll prior to ejection. If the osmotic granulation contains too much air resulting in lamination, precompression can be used. One tablet is produced for each full rotation of the press, where two are produced if running single-layer tablets.

Individual cams comprise a transport system that the punches must travel. This involves straight sections as well as curves. Working cams are usually made of bronze or very dense plastic to reduce wear on the more expensive, hardened, steel punch heads. Proper lubrication is a must to extend tool life.

Punches are guided under compression rolls that apply the forces needed to get adequate compression and consolidation. The result is a consistent filling of the dies and the compaction of the granulation by the punches. The presses have sensors that monitor the force exerted by the compression rolls so that tablets of the correct hardness and friability are produced. The tablet presses contain all the controls and monitoring sensors needed to ensure the production of high-quality tablets.

Every set of tooling produced has a safe maximum punch tip load based on its size, shape, and the type of steel used. If excessive forces are applied, damage to the punch tips may occur, causing tableting problems and possible further damage to the press. As a result, there is an overload system, "safety valve," built into the press that detects forces above the set level and releases the lower main compression roll to relieve the force. The system uses hydraulic cylinders and an air accumulator to link the upper and lower main compression rolls. It is designed to rapidly release the pressure, which saves the punch tips from damage. However, it must be set properly, based on the tooling limits and the product being produced (the amount of pressure needed to get a quality tablet).

Monitoring system allows:

1. measuring the compaction force in the die using strain gauges on the overload system
2. monitoring pressures within the accumulator. This provides a much better picture of what is happening to the granulation during compression. The overload setting should never be over the maximum allowable load, yet it should be high enough so that it doesn't release at normal consolidation force levels. Precompression rolls are also protected by a fixed (2,000 Ibs) overload safety release mechanism.

Tablet thickness is established by the main compression rolls. They do not move except in an overload situation. As a result, the punch tips of a pair of punches (upper and lower) always end up the same distance apart. There is some expansion of the tablet after release of the consolidation force due to the elastic rebound of its components. The weight of the tablet is based on the volume of granulation in the die cavity.

The hardness is determined by the amount of pressure applied, which at a given tablet thickness is also determined by the fill. Hence, at a given volume, if the main compression rolls are moved closer together, the pressure will be increased, and the tablet will be thinner and harder. If the compression rolls remain at the same distance and the volume of granulation is decreased (possibly due to trapped air pockets), the tablet will be not only underweight but not as hard and more friable. Further, if there are variations in the total length of punch pairs (upper and lower), within the set of 55 pairs of punches, it will also result in variations in tablet thickness, hardness, and friability among the 55 tablets produced in a cycle.

Another consideration in tableting is obtaining a tablet strong enough to hold up under further manufacturing processes, and porous enough to dissolve/break up when water enters the core. If the tablet is compacted at too high a pressure or with too long a dwell time, the core could become so dense that water cannot easily penetrate.

When press speed increases, the machine performs all its functions at an accelerated rate. This will decrease the time in which each function can be performed. Time under the feeder will decrease, so to compensate, changes in flow rate, paddle speed, and possibly fill cam may be required. Otherwise, inconsistent tablet weights may result. Compression dwell times will also decrease, air entrapment may increase, and elastic properties of granulation may change, causing lamination and capping. Adjustments to precompression, main compression, and punch entry may be required in addition to overload set points. Ejection forces will also increase.

The key to getting high-quality tablets is

1. consistent die fill (volume/weight)
2. enough force to compact the granulation into a tablet with desired hardness and friability
3. sets of punches with the same length per pair
4. smooth operating punches
5. non-sticky granulation.

CORE COMPRESSION MANUFACTURING PROCESS

Line Preparation and Sign-Off

Authorization to proceed must be obtained. This is mainly to ensure that similar size/shape tablets were not produced on the press in the previous batch, and to also ensure that previous processes are complete and there are no problems with the granulation – Line Clearance.

Tooling check-in is accomplished to ensure proper size/shape punches and dies for the product and that they are in good condition and properly lubricated. Set up tote bin(s) of granulation at dump station(s) and establish a good seal between the collar and tote bottom(s). Set up presses based on MF parameters for start-up. For bi-layer tablets, adjust the precompression roll force (position) to obtain the required thickness of the active drug layer. Perform in-process testing and choose the optimum precompression force that yields the desired layer weight. Add the osmotic granulation layer and test for total tablet weight, hardness, thickness, and friability. Adjust controls as necessary to obtain MF specifications.

Following start-up, perform in-process testing per SOP and MF:

1. Average hardness based on initial 10 core sample
2. Average active (drug) layer weight based on the sample of 55 cores. Average total core weight based on the sample of 55 cores. Weight variation
3. Average friability based on the sample of 55 cores
4. Average thickness based on initial 10 core sample
5. Coefficient of variation (CV), a statistical measure of tablet variation
6. Conduct friability testing per SOP record values
7. Perform visual inspection AQLs per SOP and record on appropriate production forms
8. Pull samples per SOP for assay lab testing
9. Weigh acceptable tablets and store them in appropriately labeled containers
10. Perform accountability and calculate theoretical yield
11. Move tablets to staging area for sub-coating or membrane coating

Laboratory Analysis of Cores

The samples of cores are sent to the lab for assay analysis. They check for drug content, purity, and uniformity.

Critical Elements in Tablet Compression:

1. Granulation has good flow and compaction qualities and proper moisture levels
2. Proper assembly of the press prior to start-up includes the following:
   The right toolset of punches and dies
   The right cams and their proper placement and adjustment
   Having the feed frame perfectly level with the turret die table and die
   Proper lubrication of punches
   Clean, polished punch bores that allow easy move punches
   Dies seated flush with the turret's die table and die lock screws torqued to proper specifications
   Properly set overload mechanism
3. Closely follow start-up procedures to ensure that the press is correctly set. One rotation by hand should be performed to detect improper set-up conditions.
4. Monitor the press for smooth movement of the punches as they ride the cams.
5. Listen for unusual noises that may indicate problems with the press.
6. Proper vacuuming action in the tableting area removes dust generated by the movement of the punches but is not so great that it draws granulation from the dies.

Tableting Defects

1. Tablets stick to upper punches
2. Safety overload release operating
3. Lower punch binding in die
4. Excess machine vibration
5. Tablet breaking on take-off
6. Non-uniform tablet weight
7. Non-uniform tablet thickness
8. Non-uniform density (friability)
9. Tablets binding in dies
10. Excess punch head or cam wear
11. Loss of material
12. Granulation leakage
13. Dirt in product (black specks/grease)
14. Delamination (bi-layer only)/lamination (single layer only)
15. Picking (chips in surface of tablet)
16. Flashing (crowning on tablet edge)
17. Capping (top of tablet comes off)

Trouble-Shooting Guide (After some types of maintenance recalibration may be required)

Tablets Stick to Upper Punches

Damaged upper punch face	Replace/refurbish punches
High moisture content in granulation	Re-dry granulation – check room humidity
Insufficient lubricant	Increase lubricant
Pressure roll bounce	Overload not properly set
Insufficient compaction force	Reduce weight increase thickness within tolerance

Safety Overload Release Operating

Damaged punches (burr on tip)	Repair/replace
Excessive pressure	Reduce pressure (increase thickness or weight
Overload pressure set too close	Reset overload within limits of tableting pressure

Lower Punch Binding in Die

Material stuck to punch/die	Check for worn punch/die
High moisture level in granulation	Dry granulation

Excess Machine Vibration

Worn drive belt	Inspect and replace
Mismatched punches	Repair or replace
Operating near density point granulation	Increase thickness or lower weight
Heavy/excessive ejection pressure	Barreled dies (wear band in center of die)
	Worn ejection cam
Incorrect overload pressure release setting	Increase load setting (within tool limits)

Tablet Breaking on Take-Off

Take-off plate incorrectly set	Check and reset
Lower punch ejection height incorrect	Check and reset
Insufficient load at compression point	Increase pressure and check overload
Speed too high for product	Reduce speed until acceptable
Not enough binder in formulation	Increase binder in formulation
Granules too dry	Increase moisture level
Insufficient feed to dies (soft tablets)	Check hopper-flow

Non-uniform Tablet Weight

Erratic punch flight	Free punches
Excess vibration	Worn/loose weight adjustment
Lower punch control operation	Limit cam worn
Material loss/gain after die fill	"Tail over die," if installed, not flat
Re-circulating band leaking	Excess cleaning vacuum
Feed frame starved or choked	Wrong hopper adjustment
Bridging in hopper	Wrong fill cam
Too much re-circulation	Die not filling
Press running too fast or feeder too slow	Feed frame choked
Bad scrape off	Damaged blade or bad spring
Non-uniform punch length	Punch length within $\pm 0.001''$
Die projecting above die table	Clean die pocket and reseat die
Inconsistent granules	Re-mill granulation

Non-uniform Tablet Thickness

Material loss/gain after die fill	Feed frame starved or choked
Pressure roll bounce	Overload release incorrectly set

Non-uniform Density (Friability)

Uneven granule distribution in die	Layering or segregation/separation of components

Tablets Binding in Dies

Worn dies (barreled/rough surface)	Reverse die or repair/replace
High moisture level in granulation	Dry granulation

Excess Punch Head or Cam Wear

Binding punches	Dirty punch guides
Rough spots from previous operations	Polish/replace

Loss of Material

Incorrect feed frame-to-die-table setup	Feed frame is not level or is worn.
Incorrect action on re-circulating band	Gap between bottom and die table
Die table scraper action insufficient	Scraper blade worn/bent
Loss at compression point	Compressing too high in the die
Lower punch penetration set too high	Check and reset
Worn lower punches or dies	Replace
Recirculating channel worn/missing	Replace
Excessive material flow to feed frame	Reduce flow rate

Granulation Leakage

Die table run-out	Incorrect clearance between feed frame/table
Cam problems	Pull down cams enough to minimize
Overfilling feed frame	Feed frame correctly set to supply granulation

Dirt in Product (Black Specks)

Extruded granulation from punch bores	Build-up in punch bores
Osmotic mixing with active granulation	Scraper blades damaged

Delamination (Bi-layer Only)/Lamination (Single Layer Only)

Insufficient compaction	Insufficient feed to dies low pressure Trapped air – increase precompression
Bi-layer separation	Excess precompression force

Picking (Chips Out of Surface of Tablet)

Excess moisture	Moisture level of granulation too high High humidity in compression area
Ejection problems (sticking)	Insufficient lubrication in granulation Rough/pitted die walls Barreled dies (worn more in center)

4 Sub-coating

Sub-coating is applied to the tablet core prior to the semi-permeable membrane. It establishes the tablet's 4–5 hours delayed release. Coating is an aqueous (water-soluble) polymer-based film coating. Type of components used in the coating, its thickness, and the fact that it is inside the semi-permeable membrane determine its rate of dissolution and delay time. The successful application of a film coating to tablets depends on several interrelated factors. The ideal film-coating process should deposit a given quantity of solids as a cohesive unit, as evenly as possible and in the shortest possible time, onto a batch of tablets. To achieve this, the coating system would

1. dry the coating at the same rate at which it was applied, so that the tablets don't stick together.
2. evenly cover the surface of the tablets at the rate at which new surfaces are exposed, so that the covering is smooth and has a uniform thickness.

To apply the sub-coating, e.g., a Vector-Freund Hi-Coater, with a 67″ stretched bed that allows a full lot to be sub-coated at a time could be utilized. The coater has a somewhat angular round pan that rotates on a horizontal axis. The inside of the pan is fitted with mixing baffles. The film-forming solution is sprayed onto the tablets as the pan rotates (continuously mixing and moving the tablets). Heated air is drawn through the bed of tablets to dry the film solution just after it hits the surface of the tablets. Major functions that govern the process are

1. tablet mixing ability of the coating pan
2. drying ability of the coater

SUB-COATING MANUFACTURING PROCESS

Following line clearance. Receive and verify cores from compression and film-forming materials.

Start circulation of hot water in vessel jacket. Dispense specified amount of water into tank and set mixer to 110 rpm. After it reaches 55°C, increase mixer speed to 1,145 rpm (1,120–1,170). Add dry ingredients in the specified order while maintaining 55°C. Mix at 1,145 rpm until all solids are dissolved/hydrated (minimum of 2 hours).

Reduce mixer speed to 110 rpm and continue throughout sub-coating while keeping it at 55°C. Temperature has a significant effect on viscosity of solution during atomization, adhesion, and coalescence. 55°C setting also helps to inhibit microbial growth.

Check gun flow rate prior to loading cores. This is critical to ensure that guns are set up and operating properly. Set up coater controls per MF parameters. Sub-coat tablets using parameters in MF per SOP.

DOI: 10.1201/9781003224716-4

Pre-heat coater pan. Load cores and rotate pan 1–2 revolutions. Determine average initial weight of 300 uncoated cores. Insert air deflector and pre-heat loaded pan to target temperature. Adjust pan speed to 5 rpm. Adjust spray rate and pan speed at 60 and at 120 minutes into the spray cycle per MF.

Dry at 7 rpm with exhaust set at 45°C and air volume at 3,500 cfm. Dry for 120 minutes (120–180).

Remove cores from coater, by opening the pan's trap door and dumping them into a tablet bin. Then place the coated cores in tared polybag-lined plastic barrels and weigh. Sample and perform AQLs per SOPs. Perform accountability and move cores to membrane coating staging area.

CRITICAL ELEMENTS AND TROUBLE SHOOTING

1. To achieve a film coating with the proper delay characteristics, key factors are:
   Inlet air humidity
   Exhaust air temperature
   Air flow (volume) to the pan (slight negative pressure in pan of −1″ to −2″)
   Pan speed and mixing conditions within the bed
   Position and geometry of the spray guns
   Spray rate, atomization air pattern, and air volumes
2. The validated size of the load for a certain range of total core weight. Smaller bed will affect exhaust duct coverage. Larger load might impede effective coating
3. Quality of the film-forming suspension

   Temperature has a significant effect on the viscosity of this polymer while in a water suspension and helps inhibit microbial growth/contamination.

   Polymer concentration likewise has a significant effect on viscosity.

   Higher-than-specified viscosity can lead to rough (poorly coalesced) film. This semi-permeable membrane controls the osmotic pressure and dissolution of the sub-coating and drug release rates after time delay.

Trouble-Shooting Guide

Incomplete mixing	Use of low shear mixer
Temperature too high or low	Lumps in coating
Foaming of coating	Mixing at too high speed
Suspension	Improper heating
Variation in coating coverage	Uneven spray pattern
Twinning/picking/peeling	Guns too close to bed
Tablets stick together (twins)	Improperly spaced gun(s)
Over-wetting of cores	Low process air temperature
RH too high	Chiller/condenser not working
Suspension dries too much	Gun-to-bed distance too high
Cracks in coating	Poor cohesion/elasticity-low process temperature
Blocked nozzles	Poorly mixed suspension solution

5 Membrane Coating

Semi-permeable membrane coating technology is the critical factor in establishing drug-delivery rate. Membrane coating and aqueous sub-coating involve nearly identical film coating technologies.

The main difference is in the components used in membrane coating. A primary polymer that is soluble in organic solvents and insoluble in water is used in this application. This membrane needs to maintain its integrity while transiting in watery gastrointestinal tract. The intended release is to provide a higher degree of precision in delivering the drug. To obtain this controlled release action, the semi-permeable membrane acts to control osmotic pressure and desired drug delivery.

Cellulose acetate is used as the primary membrane film coating polymer. Typically, average molecular weight specifications for cellulose acetate are 320S and 398-10. Cellulose acetate 320S forms membranes with slightly larger pores, which increase the movement of water through the membrane. In addition to cellulose acetate at 60% solids, hydroxypropyl cellulose at 35% solids might be used in combination with a plasticizer at 5%. The hydroxypropyl cellulose is soluble in water and acts as a flux enhancer. After ingestion, it is dissolved by water in the digestive system and opens the membrane, increasing its permeability.

All membrane coatings include a plasticizer (which is the same as in sub-coating), polyethylene glycol 3350 (PEG 3350). In the membrane coating, it is also at a concentration of about 5% of solids by weight. PEG 3350 is soluble in water and organic solvents. Since PEG 3350 is soluble in both, it serves a dual purpose:

1. provides the elastic flexibility needed during manufacturing
2. acts as a flux enhancer, dissolving the membrane after administration, slightly enhancing permeability.

Polymers are hydrated and suspended in acetone-based solution or methylene chloride. For the acetone-based solution, start hydrating the PEG 3350 in purified water (PW), then adding acetone to it, and finally the cellulose acetate. The PW accounts for only about 5%–10% of the solvent used. For the methylene chloride solution, dispense methylene chloride and methyl alcohol in mixing tank and add the polymers in the order specified on MF. The polymers will wet adequately in these solvents, so PW does not have to be used. Methanol accounts for 10%–20% of the solvent depending on product. For both solvent mixtures, agitate the ingredients until dissolved and/or for a minimum time to allow wetting of the polymers.

Acetone is less expensive than MeCI2, and it is also much easier and cheaper to dispose of after use. MeCI2 provides a membrane that better matches the desired traits for that system. However, acetone and water or straight acetone is the preferred solvent. Polymer and Solvent Breakdown (Table 5.1).

The method, sequence of addition, and exact quantities are all very critical in obtaining a membrane with the right characteristics. The solvent ratio is critical to

DOI: 10.1201/9781003224716-5

TABLE 5.1
Membrane Coatings

Primary Polymers (% by Wt.)	Plasticizer (% by Wt.)	Solvents (% by Wt.)
Cellulose acetate (398-10) – 95%	Polyethylene glycol 3350 – 5%	MeCI + 20% MeOH – 95%
Cellulose acetate (398-10) (60%) + Hydroxypropyl cellulose (35%)	Polyethylene glycol 3350 – 5%	MeCI + 20% MeOH – 95%
Cellulose acetate (398-10) (95%)	Polyethylene glycol 3350 – 5%	Acetone + 5% H_2O – 95%

membrane coating. Other critical factors to membrane's characteristics are the spray rate and bed humidity (temperature and air volume).

Solution mass flow meters (e.g., micro-motion) on all solvent coater tanks measure the quantity of solution in kilograms that is being dispensed from the spray pump. The system is based on measuring the flowing solution rate and is temperature compensated. The micro-motion system establishes the vibration in a loop of the solution line and then detects the changes, which vary based on the motion of the solution's mass in the loop. This allows the operator to more precisely control the amount of solvent placed in the tank during mixing and the amount of solution sprayed.

The resulting film that forms on the tablets using these polymers has semi-permeable properties. Permeability is affected by varying the thickness of the membrane. The density of the membrane really establishes its permeability. However, density is not easily quantified, hence the use of weight gain as an approximation. Release rates can be problematic. Therefore, to consistently achieve the specified release rate, the membrane must be designed to withstand the hydrostatic pressure that builds up inside the core during normal dosing. Membrane should be flexible and not too brittle, as it might crack/split under the required pressures. As a result, it becomes much more important in membrane coating to get a well-mixed and evenly applied layer of film on the tablet surface than it did in aqueous sub-coating.

There are imperfections on the surface of the core that can cause membrane ruptures whereby splits could occur. This would have a significant impact on the programmed delivery rate of a well-compressed core compared to one that has "flashing," a cosmetic surface problem due to damaged or worn compression punches/dies.

The organic-solvent-based film coating processes also require some additional safeguards to prevent inhalation of toxic fumes, absorption of the solvent through the skin, explosions, and environmental damage. The process operators are required to wear supplied air respirators with Tyvek suits whenever solvent fumes could get into the room, which is primarily while dispensing solvents from drums, addition of solids, and cleaning. Blue neoprene gloves are required for handling acetone or methylene chloride, and regular latex gloves for isopropyl alcohol (IPA) are used in cleaning. The flammability and explosive potential are minimized in the following ways:

1. Sealed solution tanks that are purged with N_2 and have O_2-level sensors
2. Maintaining a negative pressure in the coater during spraying and drying
3. Special conductive floor covering on coater room floors
4. Wearing approved conductive-soled shoes or booties

5. Grounding and/or bonding together of all drums, tanks, transfer containers, pumps, spray boom, and portable equipment (pallet jacks, tablet bins, dollies, etc.)
6. Special non-sparking tools
7. Oxygen sensors that monitor safe breathing level with alarms – intermittently flashing blue lights at 20% O_2 and continuously flash blue lights at 18% O_2
8. Solvent vapor monitoring and alarm systems that detect LEL (lower explosive limit, the minimum level to sustain combustion) – intermittently flashing blue lights at 10% LEL and continuously flashing blue lights at 20% LEL
9. Both O_2 and LEL levels can be observed on the gas monitoring status panel outside room
10. The coaters' exhaust airflow is monitored for LEL, which is a function of spray rate and air volume to the pan
11. The coaters' LEL levels from two sensors can be displayed on the coaters' computerized control panel
12. When 15% of LEL in exhaust air is reached, a red warning light illuminates and a horn sounds
13. When it reaches 50% of LEL, control panel displays a warning, and the solution pumps are shutdown
14. Coaters are equipped with explosion protective systems that discharge a halo suppressant into the pan inlet and exhaust ducts before the LEL reaches an explosive level (they must be manually armed before spraying)
15. The rooms have blow-out panels in the back walls that are designed to minimize damage to the building and personnel by giving the explosion an easy way out

Organic solvents evaporate much faster at equivalent temperatures, thus increasing spray rate and reducing air volume to achieve the adhesive and cohesive film properties needed. This change in evaporation also impacts the exhaust temperature. Higher rate of evaporation results in cooler exhaust temperatures – most products specify a setting around 20°C–22°C compared to the 45°C of sub-coating. The inlet air is still run through an air chiller/condenser to reduce its humidity. Those are the major differences between aqueous film coating and membrane film coating processes.

MEMBRANE-COATING MANUFACTURING PROCESS: FOR ACETONE-BASED SOLUTIONS

Following line clearance, receive and verify cores and materials. Divide sub-coated cores into sublots of approximately equal size per MF. Set up and purge mixing tank per SOP. Prepare solution: Mix PEG (polyethylene glycol) and purified water (PW) per MF. Charge main tank with PW and/or PEG/PW solution based on MF. Maintain sealed tank(s) to prevent evaporation and introduction of O_2. Dispense acetone into tanks per SOP. Add remaining powders while solution is mixing at high-speed setting. After specified time, visually check that all solids are dissolved. If solution is not used within 24 hours, stir for 5 minutes on high before resumption of spraying.

Record mixing data in MF. Check spray rate per gun prior to use per SOP. Check and set atomization and pattern air volumes prior to loading cores per SOP and MF. Set up coater based on MF parameters. Coat following procedures in SOP and MF: Load cores (one sublot) and rotate pan 1–2 revolutions. Determine average initial weight (300-core sample). Determine the amount of solution needed. Preheat loaded pan to target exhaust temperature. Set gun-to-bed distance to MF specified height (have 2–3″ extra to adjust for bed growth while spraying). Set pan speed rpm and spray rate to specified values. Watch tablet bed for initial increase during first 10 kg of solution spray. Every 30 minutes, check/adjust gun-to-bed distance as necessary. Hourly record data and determine weight gain (100-core sample). Periodically check spray rate and gun nozzles, clean/adjust as required. Continue spraying until MF target weight gain is achieved. Operators can determine/monitor the rate of weight gain to achieve better accuracy in hitting the target value. Determine average final weight (300-core sample). Evacuate cores if problems will prevent spraying for 2+ hours per SOP. (Note: Tablet handling SOP allows max of 24 hours in dump bins.) Solution for next lot may be mixed while spraying per SOP. Set up coater for unloading. Drop cores into tablet dump bin and transfer to tarred poly-lined drums.

Accept that half-hour's increment if zero defects. If not zero defects, then segregate it and previous half-hour's increment. Perform AQL on all segregated increments. If AQL fails, then 100% inspection is required. Perform AQL on each drum. Perform accountability.

CRITICAL ELEMENTS OF COATING PROCESS

1. Solution preparation both in sequence and accuracy of weights
2. Spraying factors are critical
    - Hitting target weight gain
    - Spray rate
    - Air volume to pan
    - Exhaust temperature
    - Gun-to-bed distance
    - Partially clogged nozzle-dripping/squirting
    - Poor spray pattern/atomization (gun geometry)
3. Explosive and toxic vapor/absorption (safety issues)

Aqueous Coaters

Aqueous coaters are side vented where heated inlet air is introduced to the pan through six circular screened openings evenly spaced around the back wall of the pan. The air is then exhausted through six perforated segments, each located 60° to one another around the cylindrical region of the pan. Each of these perforated sections acts as the opening to an exhaust air duct fixed to the outside of the rotating pan. As the perforated section with its duct passes over the exhaust plenum, located between the 6 and 9 o'clock position, air is drawn from the pan using suction. The rotating bed of tablets also occupies the 6–9 o'clock position in the rotating pan.

Thus, as one of the six perforated exhaust ducts passes over the plenum, air is drawn out of the pan through the bed of tablets. The advantage of this is that airflow helps draw the sprayed solution onto the tablets and the only air leaving the pan is through the tablet bed, so there is very little solution lost in the exhaust air and very little spray pattern distortion. The exhaust air volume is purposely kept slightly higher than the inlet air volume to create a slight negative pressure inside the coater (−1 to −2″ of Hg per MF). The heated airflow is designed to dry the film-forming solution quickly enough that the tablets will not have a chance to stick together.

The mixing ability of the pan is improved using baffles (curved plates) mounted to the inner surface of the pan. The baffles ensure that all tablets come to the surface during the process by breaking up the eddies that might otherwise form in the rotating bed.

MEMBRANE-COATING MANUFACTURING PROCESS: FOR METHYLENE CHLORIDE-BASED SOLUTION

Verify line clearance

Receive and verify cores and materials

Divide sub-coated cores into sublots of approximately equal size per MF

Set up and purge mixing tank per SOPs. Prepare solution:

- Maintain sealed tank(s) to prevent evaporation and/or introduction of O_2
- Dispense solvents into mixing vessels
- Charge powders into mixing vessel while solution is mixing at high-speed setting. Visually check that all solids are dissolved and verify with a sample, if required by MF.
- If solution is not used within 24 hours, stir for 5 minutes on high before the resumption of spraying. Record mixing data in MF.

Check gun flow rates and atomization and pattern air volumes prior to loading cores per SOP. Set up coater based on MF parameters. Coat following procedures in SOP and MF:

- Load cores and rotate pan 1–2 revolutions.
- Determine average initial weight (300-core sample).
- Preheat loaded pan to target exhaust temperature.
- Set gun-to-bed distance to MF-specified height (have 2–3″ extra room to adjust for bed growth while spraying).
- Set pan speed rpm and spray rate to specified values. Watch tablet bed for initial increase during first 10 kg of solution spray. Every 30 minutes check/adjust gun-to-bed distance as necessary.
- Hourly record data and determine weight gain (100-core sample). Periodically check spray rate and nozzles, and clean/adjust as necessary. Continue spraying until target weight gain is achieved. Operators can determine/monitor the rate of weight gain to achieve better accuracy in hitting target value.
- Determine average final weight (300-core sample).
- Evacuate cores if problems will prevent spraying for 2+ hours per SOP. (Note: Tablet handling SOP allows max of 24 hours in dump bins.)

Set up coater for unloading per SOP and disarm explosion protection system. Drop cores from coater into tablet dump bin and transfer to tarred poly-lined barrels/drums.

Solution for next lot may be mixed while spraying per SOP. Pull samples, perform AQLs and accountability, and move tablets to staging.

The drying capacity of the air in coater is dependent on the temperature and humidity of the process air. The lower its relative humidity and the warmer it is, the greater its drying capacity (ability to take away water vapor). Pass process air through a chiller to condense out as much moisture as possible prior to heating it. This allows obtaining consistent and reproducible results regardless of the outside air RH. Inlet air temperature will vary to obtain a stable target exhaust temperature. Evaporation of water from the coating cools the air, so that while spraying you would expect to see a higher inlet temperature. During the drying phase, the inlet temperature will gradually fall until it equalizes with the exhaust temperature, indicating that there is very little moisture left in the tablets. Tablets and pan are heated to MF-specified temperature prior to spraying. This ensures that the coating will dry enough prior to being folded under the bed.

When tablets are in bed, ready to be dried, a film-forming solution is sprayed by a set of spray nozzles equally spaced along a spray bar that fits horizontally through the door and reaches nearly to the back of the coater pan. Proper gun placement is critical to the process and accounts for many coating problems. The gun parameters or geometry that establish the effectiveness of the spray system are

1. distance from end guns to the sides of the coater pan
2. inter spray gun distances must be equal from the front to the back of the pan
3. no crooked or cocked spray guns (parallel to each other and perpendicular to the tablet bed surface)
4. gun-to-bed distance

The first and last spray guns are set up initially so that their spray patterns reach the edges of the bed but minimize over-spray onto the sides (front/back) of the pan. Once these guns are set, the remaining guns are spaced equally between them. The goal is to obtain an even spray pattern across the entire bed. Guns should be aimed perpendicular to the tablet bed and parallel to each other. Crooked guns can lead to uneven spraying and result in over-wetting and under-wetting within the same bed. The distance from the gun tips to the surface of the rotating bed also has a major impact on the film-forming ability of the droplets. If it is too high, the patterns may overlap and/or the solution droplets could dry enough in flight that they would not bond properly to the tablet surfaces. If it is too close to the bed, you could have gaps in the spray pattern and/or too many droplets hitting a tablet, causing excess wetting.

For sub-coating, guns are typically set at 8″ (7–9″). During spraying, especially during in early stages, the gun-to-bed distance will need adjusting due to tablet swelling and growth as the tablets absorb moisture and later as the coating is applied. Finally, guns should be set perpendicular to the bed and aimed at the upper third of the bed. These parameters allow even coating application and almost complete drying over the remaining two-thirds of the moving tablet bed before it is folded under the bed of tablets.

The functional components of an air-spray gun are the nozzle or fluid cap (through which the solution is metered) and the air cap (through which compressed air is delivered to create the driving force for atomization and pattern). The maximum spray rate is based on the size of the opening in the nozzle and air-controlled needle valve that regulates the flow. Typically, the air cap fits over the nozzle and forms an annulus that allows compressed air to impinge on the stream of film-coating solution emerging from the nozzle. The volume of airflow is based on the size of the air cap opening and the pressure/volume of the atomization air. This airflow causes the coating solution to be broken up into tiny, atomized droplets. Droplet size and distribution are controlled by the atomizing air. In most pan-coating operations, the air cap also has "wing tips" that permit compressed air to be directed laterally onto the atomized spray defining the spray pattern. The size/eccentricity of the oval pattern is controlled by the air volume setting, which is adjusted via a thumb screw on the spray gun head. Total airflow to each gun is monitored on the HMI (standard liters per minute) and in psi.

A positive displacement pump is used to transfer coating solution. Two counter-rotating gears are employed to draw solution into and through the pump housing. Because the gears meet the coating solution, they must be made of non-corrosive materials, usually nylon. Gear pumps can transfer solutions with a wide range of viscosities. However, they rely on the lubricating properties of the solution to minimize wear on the gear mechanism. Each gun is supplied by a separate pump that provides individual spray rates, which could be monitored for clogged nozzles. The film-coating solution is filtered through a fine mesh screen, with openings less than one-third the size of the nozzle.

To help hydrate the polymer and obtain the correct viscosity, the solution is warmed prior to mixing, and temperature is maintained during mixing and spray cycle. Tanks have built-in load scales for controlled dispensing exact quantities of liquids. Mixing process starts by filling the tank with the required amount of purified water and heating it to the required temperature. Dry components are added and mixed instantly and quickly dispersed. Solution needs to mix until a uniform and homogeneous suspension is attained, and the polymers are fully hydrated. As the polymers hydrate, the solution turns translucent. A minimum of 4 hours of mixing is specified. The mixing needs to continue at low speed throughout the spray cycle and temperature of 55°C. Poor mixing can result in lumps that could plug spray nozzles and the filter screen.

Coaters are computer controlled and computer monitored. The control station regulates all the critical process parameters, such as the following: inlet and exhaust air temperature; total and back-end inlet air volumes; pan pressure (negative), and it in turn establishes the exhaust air volume, pan speed, overall spray rate (monitoring each nozzle), and atomization and pattern air volume (pressure) to the nozzles.

A suspension combination of a polymer (hydroxyethyl cellulose), a plasticizer (polyethylene glycol 3350 (PEG)), and purified water creates a type of solution where the polymer is hydrated and suspended in the solvent vehicle at a viscosity that allows for spraying with desired adhesive and drying characteristics.

The polymer, hydroxyethyl cellulose, is chosen for its chemical and physical characteristics. It is soluble in water, but slightly hydrophobic, which slows down its dissolution rate. This makes it a better candidate for time-delay coating compared with quick-dissolving coating polymer, hydroxypropyl methylcellulose (HPMC). In addition, its molecular structure produces a coating with good adhesion, tensile strength, and elasticity – all desirable qualities for a film coating. However, films made exclusively with amorphous polymers have good toughness and hardness but tend to be brittle. This physical property is based on the polymer glass-transition temperature, a point where it exhibits properties like those of inorganic glass. This temperature is well above temperatures used in the coating. To maintain a more elastic quality and mechanical strength, a plasticizer (PEG) is added to the film-forming mixture. This effectively reduces the glass-transition temperature and imparts some flexibility to the coating. The plasticizer is chosen for its physical and chemical compatibility with the primary polymer. Both molecular weight and concentration of the plasticizer affect the glass-transition temperature of the mixture. The lower the molecular weight and/or the higher the concentration of PEG, the greater the reduction in glass-transition temperature and consequently the more elastic, but with a lower tensile strength.

A coating that can withstand the mechanical stresses of further manufacturing and is flexible enough to allow some expansion and contraction of the cores without cracking is required. PEG, with an average molecular weight of 3,350, is in about the middle range of desirable molecular weights. With a concentration of 5% by weight, it provides a film coating that is strong and flexible to handle the stresses of the film shrinkage during drying and the slight tablet expansion and contraction in follow-on processes.

The film-forming mixture of polymer and plasticizer is blended with purified water. The resulting suspension is atomized and delivered to the surface of the cores with enough velocity and fluidity to wet the surface, spread out, and coalesce to form the film. It is imperative that the coating solution not only spread out but also dry nearly instantaneously on contact with the core. Otherwise, the cores may stick together (twinning) and possibly pull off pieces of each other under the stresses in the bed (what is called: sticking and picking). Film coating requires a delicate balance between

1. rate of coating application
2. size and velocity of the droplet
3. droplet viscosity on impact
4. drying capacity: air temperature, RH, and volume.

6 Drilling

To produce precisely sized orifices at higher speeds, lasers are used to produce, drilling up to 5 holes per tablet, up to 100,000 tablets per hour. A laser device generates, concentrates, and focuses a coherent beam of photons. Photons (light particles) are given off when an electron in an atom moves closer to the nucleus (lower energy state). The law of "conservation of energy" requires energy to go somewhere. If an electron's energy is reduced, it is converted into a moving particle called a photon. Photons can be thought of as little packets of energy that vibrate in a wave-like fashion as they move in a straight line. How quickly the photons vibrate (their frequency) determines the color/ type of light. If those small packets of energy are concentrated and focused, then they will generate heat. To get a relatively high energy level that can be accurately controlled, a coherent light source is needed that must generate enough photons to do the work. If all the photons are oscillating at the same frequency, it is much easier to get them all to go in one direction. Hence, the narrow beam we see on laser pointers. A coherent beam of photons (laser) of same wavelength, frequency, phase, and direction.

There are many different types of lasers that use different kinds of atoms to generate photons. The use of a mixture of carbon dioxide (CO_2), nitrogen (N_2), and helium (He) is one kind. The electrons in helium atoms, which make up about 80% of the gas mixture, are easily moved to a higher excitation level by an electrical charge. It passes its energy to a nitrogen atom, which makes up 15% of the gas in this kind of laser. The nitrogen atoms, through collisions, then transfer their excess energy to CO_2 molecules, 5% of the gas. Energy is stored in the molecule by moving an electron to a high state of excitation. The amount of excitation that takes place determines how high the electron goes. The excitations are random; that is, electrons are raised into many different levels. When the electron drops back to a lower excitation level it generates a photon at a specific wavelength, which is dependent on the type of atom and the excitation level: In this kind of laser, the wavelength is 10.6/1 m (millions of a meter). Most of the excited states decay rather quickly, except for those electrons in the metastable state. Eventually, this leaves many atoms with an electron in the metastable excited state and all other atoms in the ground state. When more than half the atoms have electrons in metastable states, then a "population inversion" has been achieved and the laser is ready to fire. Metastable state electrons deexcite relatively slowly by themselves, but each one emits a photon that can stimulate the deexcitation of others. It is the resulting chain reaction of "stimulated emissions" that allow the laser to reach relatively high levels of amplification.

When the "spontaneously emitted" photons collide with CO_2 molecules that have electrons at the metastable state, those molecules emit photons with identical frequency and direction (phase). There are now two identical photons, both going in the same direction, looking for even more excited CO_2 molecules. This "stimulated emission" process, where the photons are stimulating or generating additional photon

DOI: 10.1201/9781003224716-6

emissions, is the key to laser physics. If they remain contained and have a fresh supply of gases with electrons at the metastable state, the number of coherent photons will continue to increase or be amplified. The use of this amplification process based on stimulated emission is the heart of the laser technology – *Light Amplification by Stimulated Emission of Radiation* (LASER).

To help maintain this interaction by raising new electrons to their metastable state, the electrical discharge that started the process continues, but at a lower level. The optical energy at a single wavelength is contained during amplification by the laser cavity, a chamber of mirrors. This single-frequency energy is amplified, and a portion of it can pass through one of the mirrors that make up the optical cavity. In other words, it is reflected and bounced back and forth between mirrors until sufficiently amplified. The output mirrors are designed to reflect only a certain percentage of photons at a specific frequency that strike them partially transmissive mirrors (to photons at 10.6 flm). The photons that pass through the mirror from the beam are used to drill the tablets. The photons that are reflected remain in the optical cavity, which helps to sustain the stimulated emission process. As a result, energy requirements for continuous operation are considerably less than during start-up. The back mirrors are 99% reflective, letting just 1% of the photons penetrate for detection by a power meter that measures the laser's power output.

To obtain adequate amplification, the mirrors must be at a certain distance apart. This distance determines the output power of the laser. To reduce the overall length of the housing, corner blocks (beam-folding mirrors) are used to bend/fold the beam. The laser cavities are either "C" shaped or "2" shaped. Despite this, the housing for the optical cavity is still about 20 ft long. The beam passes through the output mirror and is diverted by a shutter until it is needed for drilling: The shutter is located where the beam exits the laser enclosure on its way to the drilling unit. The shutter has a reflective gold-coated blade, which is rotated into the path of the beam by a rotary solenoid. The solenoid fails to the closed position. In the closed position, the beam is deflected 90° into "beam dump" that absorbs laser energy. While the shutter is open, a warning indicator is illuminated. If the shutter fails/overheats, the laser is shutdown.

The heart of the laser is the optical cavity or resonator, which consists of glass tubes mouthed between metal blocks. The blocks hold plasma tubes, mirrors, electrodes, and gas and cooling fluid. The tubes must be perfectly parallel, and the mirrors must be perpendicular to the long axes. The resonator must be held extremely stable to maintain the alignment of mirrors and plasma tubes. This requires a precisely maintained system temperature for the entire assembly. Heat builds up rapidly within the laser, and without adequate cooling and well-designed support structures, the mirrors would be out of alignment in no time. The system temperature is regulated as close to room temperature (about 25°C) as possible with an interlock sensor that shuts down the system if it gets 5°C above the regulated temperature. The entire plasma tube is encased in a cooling tube through which cooled oil is recirculated. The oil is cooled by chilled water in a heat exchanger within the laser enclosure. A gas system provides a continuous flow of chemically pure gases in the proper concentrations (approximately 5% CO_2, 15% N_2, and 80% He). The gases are removed from the plasma tubes by a vacuum pump. The equipment has the capability to recycle the gases through a series of pressure regulators, traps, filters, and catalysts that remove wastes. Then new gases are added as needed to maintain the right mix. There is also a large power supply system that maintains a regulated high-voltage current flow to the plasma tubes and provides correct voltages for start-up, pulsing, and continuous operations. It must convert 480 V AC current into approximately 20,000 V DC current and regulate the DC voltage that flows through the plasma tube.

The drilling unit is a computer-controlled system that receives input from several detectors and sensors. The drilling unit directs the laser beam through a focusing lens onto the tablets for drilling. A separate beam is directed to each side. The beam is diverted by a mirror and corner blocks to the proper side for drilling. The last mirror before each focusing lens is a partially transmissive mirror with a built-in power sensor. If the power sensor does not detect a strong enough pulse to effectively drill through the membrane, a signal is sent to the logic computer to reject the tablet. The focusing lenses are slightly offset along the axis of the carrier. This allows the beam to be absorbed by a block of graphite material if a pulse is directed at an empty carrier – this keeps it from damaging the optics on the opposite side. Based on signals from the various sensors, the accept/reject criteria are evaluated. As the tablets are removed from the carriers, they are directed to the appropriate collection container by the logic computer, which controls a solenoid-driven reject flapper on the discharge chute. If all the criteria have been met, the tablet is sent down the delivery chute to the collection unit. If it fails any of the criteria, then the reject flapper diverts it, plus several other tablets on either side down the reject chute. The beam drills a hole at 1,500 tablets per minute. It is controlled electronically as high voltage DC power supplied to the plasma tubes is pulsed to generate a short burst of laser light. This burst is timed to arrive at the tablet when the spot we want drilled is passing before the focusing lens. The beam pulse is controlled by how long the electrical energy is supplied to the resonator's plasma tubes. The laser pulse continuously at a speed synchronized with the carrier and just let a graphite block absorb any missed shots when a tablet is missing.

Once a tablet is drilled, the compressed-air jets dislodge the tablets from the carrier, and they fall down the discharge chute. A solenoid-operated reject flapper-on command from the logic computer changes the path of the falling tablet so that it ends up in the reject box.

The color detector (monochromatic sensor) is used primarily to determine the side to be drilled (light/dark side) on bi-layer tablets. The hole is drilled on the side containing the drug reservoir. This is the main reason ferric oxide coloring (black/red) is added to the osmotic granulation. When set to drill bi-layer tablets, the sensor signals the logic computer to reject what it perceives to be single-layer tablets (monochrome). Therefore, it is so important to prevent osmotic granulation contamination on the active layer during tablet compression.

The adjustable parameters that affect hole size, placement, and roundness are

1. pulse width (roundness)
2. plasma current (output power orifice depth)
3. beam control (focus diameter of orifice)
4. orifice position
5. orifice spacing on multi-drilled tablets.

IN-PROCESS TESTING

In-process testing takes place at start-up and every 1/2 hour throughout the run. These are the basic tests:

1. measuring the drilled orifice on the microscope and recording the length in mm of the major and minor axes of the orifice
2. determining the aspect ratio (roundness) of the orifice
3. inspecting the appearance of the orifice
4. examining the tablet/orifice for defects
5. measuring distances between orifices on multiple-hole tablets
6. measuring penetration of the hole through the membrane and sub-coating

The drilled tablets are segregated in half-hour increments and are not added to drums until they have passed all tests and AQLs. Operators use sequentially numbered poly bags to hold the increments. Orifice measurements and examination are accomplished by using a razor blade to slice off the portion of the membrane coating in which the orifice was drilled. The aspect ratio is computed by dividing the major axis (longest) by the minor axis (shortest). Most products specify a range of 1.0. It is common practice to start adjusting laser pulse width before it gets close to the maximum. In addition to the microscopic examination, 500 tablets are inspected visually for critical defects.

Test procedures for multiple orifices on the same side of a tablet measure the distance between the holes. For a sub-coating, the membrane coating is peeled off the sub-coating, and measurements and inspection are accomplished on the sub-coating – the laser must drill through both layers.

Inspection Covers

1. foreign product/odor
2. incorrect size/color
3. gross foreign material
4. super holes (long pulses)
5. no orifice
6. single-layer tablet in bi-layer lot

Major Defects Are the Following

1. flaking membrane
2. erosion of membrane layer
3. more orifices than required
4. absence of one orifice required in a multi-orifice tablet
5. two orifices off-center
6. embedded surface spots
7. splitting
8. cracks

Laboratory Analysis: Normally, no lab analysis is accomplished after drilling. A major part of the analysis is for residual solvents, and these will not be removed until drying.

Safety Issues

1. The laser beam will reflect off nearly any metal surface, even dull ones.
2. The 10.6 ~m wavelength laser beam is invisible. Wear proper light-filtering safety glasses made entirely of plastic. They protect against only dispersed radiation, not a direct beam.
3. Use proper lock-out/tag-out procedures because of the extremely high voltage used in laser enclosure.

CRITICAL ELEMENTS OF THE DRILLING PROCESS

Drilling an orifice through the membrane is critical in obtaining the proper programmed function of the system; however, it is much easier to see the results of drilling than it is to see the results of membrane coating by using a microscope.

1. Good sorting, if required, prior to drilling helps prevent process interruptions due to twins, too thick, and under-sized tablets.
2. Start-up is critical adjusting the laser to get the right diameter circular orifice through the required coatings in the correct location.
3. Proper segregation of increments in-process testing, using microscope.
4. Challenging sensors correctly and at the proper intervals.
5. Closing shutter, if drilling is interrupted for any period.
6. Maintain lasers in peak condition, to include vacuum pump, gas mixture, controller, and sensors.

7 Drying

Following laser drilling, the cores may contain potentially harmful solvents remaining from the membrane-coating processes. To extract these solvents, heat-controlled ovens are used. Air flow, humidity, and time are monitored. Solvents are extracted from the drilled cores in the least amount of time, ensuring that the residuals are well below safe levels.

Solvents or moisture can be either chemically or physically "bound." Chemical bonding occurs most often in crystals that hold solvent or water within their crystalline structure by electrostatic force. This liquid can be removed, but often only at high temperatures and usually by destroying the crystals. Physical bonding is most often the result of surface tension, which counteracts vapor pressure to hold the liquid to the material. The granular structure of cores holds moisture in its lattice-like structure through surface tension really well. To overcome this physical bonding and remove the solvents, it would take very high temperatures and low relative humidity.

Water and organic solvent can be bound to cores material. Most of the "free" solvent in the membrane is removed during membrane coating by the warm dry process air flow. More "free" liquid evaporates from the membrane as the cores are rotated in the pan during the final weight gain measurements and during transfer to barrels. Further heating will remove some of the bound moisture, but because components are often temperature sensitive, there are limits. In addition, most of the residual solvent must be extracted through the orifice since the dried membrane acts as a barrier. In order to get residual solvents down to acceptable levels at reasonable temperatures, a mass transfer "humidity drying" mechanism is utilized. The process utilizes special temperature- and humidity-controlled ovens. These ovens are designed not only to control temperatures but also to regulate the humidity and airflow.

Tablets are removed from their drums and placed in 2″-deep trays, perforated, stainless steel. Trays are then put into a wheeled rack that can be rolled into the oven. This allows the air to freely circulate around the cores. The drying process involves a pre-heat period where the ovens and cores are warmed to 40°C–50°C (MF-specified) for approximately 5 hours at ambient relative humidity. Then water vapor is added to the airflow through the oven to maintain a relative humidity of 45%–50% (MF-specified). This warm humid condition is why it's called "humidity drying." Depending on the product and size of the tablet, batch target durations of 36–120 hours. The "humidity drying" is followed by a post-humidity drying period at the same temperature, but with ambient humidity for 2–4 hours.

The pre-heat interval accomplishes two important actions. It heats the surfaces of the oven, tablet trays, and tablets to the process temperature. This ensures that when water vapor is introduced to raise the humidity, it will not condense on the tablet or the relatively colder oven surfaces and drip down on the tablets, which might result in disfiguring and possibly ruining them. Second is the removal of any remaining "free" solvent from the membrane. During this period, the air supplied to the ovens is at

DOI: 10.1201/9781003224716-7

ambient humidity. As temperatures increase, the air will hold more moisture, hence its relative humidity will decrease. This increase in temperature and decrease in relative humidity is what helps to remove any remaining "free" moisture from the system.

Heat at 40°C–50°C, even with low humidity, will not extract enough of the residual solvents. Therefore, the use of "humidity drying" cycle as a mass-transfer mechanism to bring the residual solvent levels down to an acceptable range is recommended. Water vapor enters the core of the tablet and displaces the organic solvents. Water has a much higher surface tension than solvents to be removed. As a result, it fills in the pores within the granules, displacing the "bound" organic solvents and allowing them to evaporate. In addition, since water has a higher vapor pressure than acetone, methylene chloride, ethanol, or methanol, this also helps in the mass-transfer mechanism. Water vapor makes the "bound" organic solvent molecules "free" to evaporate by trading places with them.

Final post-humidity drying period is used to remove the "free" water and organic solvents from the tablets at the end of "humidity drying." This improves their storage potential by reducing the risk of excess moisture condensing out while the tablets are stored, awaiting overcoating and printing.

LABORATORY ANALYSIS OF CORES

All products are tested for residual solvents, water content, drug release rates, and content uniformity:

a. identity
b. assay
c. water
d. residual solvents

8 Overcoating

In some products, the overcoat also has a dose of the drug incorporated into the overcoating as a first layer: This readily soluble layer of drug provides an initial dose prior to the system's programmed release of drug. Overcoating is an aqueous polymer film coating. Membrane coaters have been used to apply aqueous overcoats when a single sub lot needs coating. In sub-coating, a blended polymer film-coating solution using purified hydroxyethyl cellulose (primary polymer) and polyethylene glycol 3,350 (plasticizer) is used. Main ingredients in overcoating are:

1. Hydroxypropyl methylcellulose (HPMC) – polymer
2. Hydroxypropyl cellulose (HPC) – polymer
3. Polyethylene glycol (PEG) – polymer/plasticizer
4. Titanium dioxide – (white opaque pigment, alga base for other transparent pigments/dyes)
5. Polysorbate 80 – emulsifier/stabilizer for the pigments/dyes (used in some blends to ensure even distribution of coloring)
6. Pigments – usually iron oxide based (used in some blends)
7. Dyes – usually FD&C aluminum lakes (transparent dyes used in some blends)

HPMC, HPC, and PEG are standard polymer film-coating components. Different molecular weights are chosen to obtain the desired film properties, which vary based on other additives and applications. The addition of large amounts of pigments can affect film strength and elasticity. Titanium dioxide is a white pigment that is added for opaqueness and/or as a base for other transparent colorants.

Once the solids are added and after the MF-specified mixing time, mixer speed is reduced. Once mixed with water, polymers require 45–60 minutes to fully hydrate at specified processing temperatures. Low-speed agitation continues during the MF-specified hydration period. Mixer speed during hydration needs to be low enough that air will not be drawn into the solution, resulting in foaming. Suspensions have solids content of 10%–15% by weight.

Overcoating parameters for spraying are very similar to sub-coating with a couple of exceptions. Pan speeds are slower: 4–4.5 rpm versus 7 rpm in sub-coating. The solution spray rate for overcoat is also lower depending on product. Decreased spray rate and pan rotation is dependent on the following:

1. **Tablet Surface** – the sub-coat is applied onto a newly compressed core, while the overcoat is being layered onto the membrane coating, which is a smoother, less porous surface
2. the less viscous nature of the solution made with HPMC as the primary polymer
3. the solution's adhesive/cohesive properties and drying properties are also different.

DOI: 10.1201/9781003224716-8

Spraying continues until the specified weight gain is reached. The overcoat is a thin layer – just thick enough to cover the membrane and provide a smooth, uniform, colored surface for printing. The operators can adjust the weight gain range to achieve adequate coverage of the cores. The sub-coating that is needed to achieve the 4–5 hour time delay has a target undried weight gain of 90 mg/tablet – more than four times that of its overcoat, approximately 20 mg/tablet.

In some applications, a drug overcoat layer is blended using the polymer, HPMC, the plasticizer, PEG, with the drug. After the addition of the polymers, the solution is heated using water jacket on the tank. When MF-specified temperature is reached, the drug API is added and dissolved. The solution can cool prior to spraying so that it is well below the thermal gel temperature of the polymers, a point where they become very viscous. The drug makes up about 80% of the solids in the solution with the polymers helping to form an adhesive film when sprayed on the cores. As a result, the film coating is not very strong and allows tastes of the drug. For example, a 150 mg tablet, spraying of the drug layer continues until an undried tablet weight gain of approximately 36 mg (contains a 29 mg dose of API). Immediately after applying the drug coating, while the cores are still in the coater, apply a colored, taste-mask overcoat with a solution made from Opadry. (Opadry is a customized, one-step film coating system which combines polymer, plasticizer, and pigment, as required, in a dry concentrate. Opadry systems can be easily dispersed in aqueous or organic solvents.) The target undried weight gain for the taste-mask layer is 20 mg.

For many products, the final step in the coating process prior to dumping the cores from the coater is to apply a very small amount of polyethylene glycol 8000 (PEG 8000). The quantity is 0.00004 g per tablet, which usually amounts to less than 50 g per coater load (500 kg). The PEG 8000 acts as a lubricant, facilitating the flow of tablets, primarily in the printing machines. If too much PEG 8000 is added, this can end up with tablets that have ink adhesion problems during printing. At the conclusion of the process, a readily soluble film coating that is tasteless, smooth, slippery when wet (easy to swallow), hard yet slightly elastic, opaque, and with a little dry lubricant is the kind of overcoat optimized on a tablet heading to the printers.

OVERCOATING

Add PEG to PW and mix until clear (minimum as specified in MF). Slowly add HPMC, mixing until clear (minimum as specified in MF) Using hot water jacket, heat solution to MF specified temperature. Slowly add drug and mix until dissolved (minimum per MF). Reduce mixer speed.

Discontinue heating per MF. Check gun flow rate prior to loading cores per SOP. Set up coater based on MF parameters. Overcoat tablets using parameters in MF per SOP. Load cores and rotate pan 1–2 revolutions. Determine average initial weight using 300 cores. Preheat loaded pan to target exhaust temperature. Adjust pan speed to specified rpm and start spraying. Increase spray rate at intervals specified in MF. Hourly, record data and determine weight gain for 100-core sample. Periodically check nozzles, adjust as necessary, and clean at least each hour. Periodically check gun-to-bed distance, adjusting as necessary. Evacuate cores if unable to spray for 2+ hours per SOP. Stop spraying cores when specified target weight gain is reached. Determine average final weight using 300 cores.

COLOR OVERCOATING

Prepare color/taste-mask overcoat suspension. Prepare color overcoat solution. Dispense specified amount of PW into vessel. Set mixer to maintain vortex. Add Opadry mix until homogenous (no lumps) and/or the minimum time as specified in MF. Periodically scrape bottom of vessel while mixing. Reduce mixer speed to low per MF. Set up coater based on MF parameters.

Overcoat tablets using parameters in MF per SOP. Determine average initial weight using 300 cores. Maintain exhaust temperature specified in MF. Set pan speed to MF-specified rpm and start spraying at MF rates. Increase spray rate at MF-specified intervals. Hourly record data and determine weight gain of 100-core sample. Periodically check nozzles, adjust as necessary, and clean at least each hour. Periodically check gun-to-bed distance, adjusting as necessary. Evacuate cores if unable to spray for 2+ hours per SOP. Stop spraying cores when specified target weight gain is achieved and the tablets are adequately coated (smooth, evenly colored, opaque coating). Determine average final weight using 300 cores.

LABORATORY ANALYSIS

Products are assay tested for purity, content uniformity, and BHT (butylated hydroxytoluene) levels, if included. These products also receive drug release-rate testing.

CRITICAL ELEMENTS AND TROUBLE SHOOTING

All the process parameters are critical because of their interdependence. Care must be taken to ensure that established targets are met as closely as possible to achieve a film coating with the proper characteristics. Key factors are (Table 8.1):

1. Inlet air humidity controlled by the air chiller/condenser impacts drying capacity, adhesion (coating-to-core/ core-to-core (twinning)/core-to-pan), and cohesion (surface quality).
2. Exhaust air temperature, which drives the inlet air temperature due to evaporative cooling and has a major effect on drying capacity of process air, adhesion, and cohesion.
3. Air volume to the pan (slight negative pressure in pan of -1″–2″ of H_2O).
4. Pan speed and mixing conditions within the bed.
5. Position and geometry of the spray guns.
6. Spray rate, atomization air, and pattern air volumes, correct nozzle, and air cap.

TABLE 8.1
Trouble-Shooting Guide – Overcoating

Incomplete mixing	Use of low shear mixer
	Improper position of blades
	Speed too low
	Mix time too short
	Temperature too high/low
Lumps in coating suspension	Mix time too short
	Added solids too fast
	Mixer speed too low
Foaming of coating	Mixing at too high a speed
Suspension	Spraying too close to high-speed mixing
Twinning/picking/peeling	Guns too close to bed
Coating cover/color variation	Uneven spray pattern
	Cocked or improperly spaced gun(s)
	Incomplete mixing
	Baffles not working properly
Cracks in coating	Poor cohesion/elasticity
	Low process temperature, drying too fast, or low PEG concentration

9 Printing

The FDA requires pharmaceutical companies to mark orally ingested solid dosage products for accurate recognition of the product. The markings can be as simple as an alpha-numeric code. The goal is for them to be discernible to patients by way of markings, which, in conjunction with color and shape, make them unique, in the hopes of reducing mistaken ingestion. For compressed cores that are not overcoated, the easiest and cheapest method is to include embossing or debossing on the tableting punches used in compression. This results in indented or raised marking on the compressed core. With current polymer film-coating technology, it is still possible to use this method even for coated tablets, as the films are thin and adhesive enough to allow the marks to be seen even after overcoating.

To avoid a possible impact on membrane strength/permeability, it is recommended to keep cores smooth and print the markings on the overcoated tablets. The use of an off-set printing process in which an inked impression from an engraved plate is first made on a rubber-blanketed cylinder, then the impression is transferred to the tablet. The engraved plate is an etched rotogravure plate (gravure roller). The product logo is etched into this very smooth stainless-steel roller. The depth of the etching determines the amount of ink picked up by the roller. A doctor blade removes excess ink from the roller leaving just the ink in the etching. The ink in the etching is then transferred to a rubber roller that has a circumference identical to the gravure roller. The ink is then transferred to the tablet surface as they are pinched between the rubber roller and the carrier ramp.

Primary printing machines such as Ackley Adjustable Incline Cantilever Ramp Printers: The tablets are fed from a stainless-steel hopper by variable tablet-feed mechanism at a controlled rate into the printer's small inlet hopper. The inlet hopper is positioned over the lower end of the incline ramp. The inlet hopper transfers the tablets onto the carrier bars, which have machined pockets designed to fit exactly the product to be printed. There are two rotating nylon brushes that can be used to help fill the pockets and ensure that only one tablet per pocket is transported.

Printers can print from 25,000 to 250,000 tablets per hour. They are powered by 120V AC power and require compressed air of 80 psi @ 20 cfm. Ackley Incline Ramp Printers or Ackley Drum Feed Printers can be used, depending on product. The drum feed printers use a metal drum with pockets to transfer tablets from the inlet hopper situated above the drum to the carriers below, instead of an incline ramp. Both prints using the same offset process. The drum feed printers can also be set up/modified to print on both sides of a tablet.

The carrier bar pockets are filled by the action of the nylon brushes and gravity. The angle of the incline ramp on which the hopper is attached can be adjusted (0°–30°) to facilitate filling the carrier pockets. A "blow back" system of compressed-air jets at the top of the inlet hopper is used to keep the carrier pockets empty until the operator is ready to start printing. The small amount of PEG 8000 that was added to most

DOI: 10.1201/9781003224716-9

of products at the end of overcoating facilitates tablet movement and seating in the carrier pockets. The ink used to mark the tablets, like the overcoating concentrate, is made and supplied by Colorcon. The trade name is Opacode, and it can be obtained in many colors. The color is based on aesthetic contrast with the overcoating, FDA ink-component requirements, and marketing considerations. The ink must be capable of producing a distinct and durable image on the product and must be compatible with the tablet surface and printing press being used.

The ink film is very similar to tablet film coatings. It consists of polymers, plasticizers, pigments/dyes, defoamers, surfactants, and solvents required to dissolve or suspend the solid components. The polymer used is a food-grade shellac or a cellulose derivative. The coloring comes from either FD&C aluminum lakes or synthetic iron-oxide-based pigments, as in overcoating. The solvents are chosen to meet two printing requirements:

1. an ability to wet the surface of the tablet and produce good adhesion
2. to control drying time on the printing equipment being used

The solvents used in the ink are probably the most critical element of the formulation. Wetting the surface means the solvents in the ink slightly dissolve the overcoating of the tablet. This is the key to getting good adhesion. The ink's formula is based primarily on the type of surface, and since all printed tablets receive an aqueous overcoat, the inks are standard. Waxed, gelatin, or enteric-coated tablets would require a different formula. Most enteric coatings work by presenting a coated surface that is stable at the highly acidic pH found in the stomach but breaks down rapidly at a less acidic (relatively more basic) pH. Materials used for enteric coatings include CAP, CAT, PVAP and HPMCP, fatty acids, waxes, shellac, plastics, and plant fibers. Cellulose acetate phthalate (CAP), also known as cellacefate, is a commonly used polymer phthalate in the formulation of pharmaceuticals, such as the enteric coating of tablets or capsules and for controlled release formulations.

The ink is a suspension and needs to be mixed well before use and periodically during printing. It is placed in a tray that can be moved up or down to immerse the lower edge of the gravure roller. The ink requires dilution from time to time due to evaporation of the solvent while in the pan. The type of solvent used, and the amount can have a significant effect on the print quality and drying time. The solvents are chosen based on MF and drying time differences. The following list gives the relative evaporative rate, fastest to slowest: ethyl alcohol, isopropyl alcohol, butyl alcohol, and propylene glycol. Methyl alcohol is not used.

Once the ink is at the right consistency, lower the gravure roller's lower edge into the ink pan so that the etched roller is coated with ink. Then the doctor blade wipes the gravure roller to clean off all the ink except for that in the etchings. The doctor blade needs to have a smooth level edge without nicks. It is held against the gravure roller by air pressure (normally set to approximately 30 psi). Once the doctor blade is in place and the gravure roller shows no ink leaks or trace lines, pressure should be reduced to the lowest possible setting that provides a clean wipe of ink from the roller. Lower pressures greatly reduce gravure roller and doctor blade wear. The doctor blade is clamped into its holder with a backup blade to add stiffness. The doctor blade is the thinner and wider of the two blades (doctor blade = 0.005″ × 0.75″ × 6.25″ and backup blade = 0.010″ × 0.6875″ × 6.25″). The backup blade is on top with the doctor blade extending about 1/16″ beyond it. Debris can catch on the doctor blade, causing ink streaking and possible gravure roller damage.

The ink that remains in the etchings on the gravure roller is transferred to the rubber roller as they are pressed together. The depth of the etching determines the amount of ink picked up by the roller. Dried ink in the etching can reduce the amount of ink transferred to the rubber roller. The ink is then transferred from the rubber roller to the tablets as it rolls over them. If the ink does not get transferred to the tablets, it partially dries on the rubber roller and is re-inked as it contacts the gravure roller. This can result in double or smudged print on the next tablet. During start-up, the first set of tablets almost always end up with double print until the carriers are filling consistently, and the tablets are picking up the ink. The operators use a vacuum to remove these tablets from the carriers. If the tablets stick to the rubber roller, then a stripper plate drops them back onto the carrier, keeping them from being pinched between the rollers.

The position of the logo on the tablets is called its registration. It can be adjusted both laterally (side-to-side) and advance-retard (front-to-back). The operators adjust the registration during set-up by using the transfer method. They place food-grade lubricant on a tablet in a carrier in front of the rubber roller and hand crank the printer, allowing the lubricant to be picked up by the rubber roller and transferred to the gravure roller. The transferred lubricant spot should encircle the etched logo.

After the ink is transferred to the tablets, they continue along the ramp so that they are nearly dry by the time they are removed from the carrier pockets by jets of compressed air. The compressed air removes them from the carrier pockets, allowing them to fall down the drop chute into an in-process bag. During printing, the operators pull 100–200 tablet samples periodically and place them in a bag labeled with the current drum number. AQL is performed from the bag of random samples. Tablets are inspected for defects. IPA is used to clean rubber roller and etchings. Butyl alcohol should not be used, especially on the rubber roller, as it will soften the rubber. The key to quality printing (Table 9.1):

1. have the ink at the right consistency
2. properly set doctor blade that is smooth/sharp
3. well-matched rollers with proper tension between them and between the rubber roller and tablets
4. proper registration (alignment of logo on the tablets)
5. consistent fill of carrier pockets
6. maintain the rollers in a clean condition (etchings and rubber surface). Tablet defects, such as chipped or rough overcoating, can easily cause difficulty in printing

TABLE 9.1
Trouble-Shooting Guide: Printing

Problem	Possible Cause
Tablets will not feed/fill	Change angle of cantilever ramp
	Lubricate tablets with PEG 8000
	Lubricate hopper with food-grade spray lubricant
Print off-center	Adjust print registration (Advance/retard or lateral)
Smudged or streaked print	Ink viscosity too high (thin it with approved solvent)
Broken incomplete print	Ink drying too fast (increase viscosity or replace ink)
Etching on gravure roller clogged/damaged	Pigment settling in trough
Debris on transfer roller or damaged transfer roller	Tension between transfer roller and tablets too low
Print weak thin line	Ink viscosity too high (thin it with approved solvent)
Squashed/broad print	Bad/worn rubber transfer roller
	Tension between transfer roller and tablets too high
Uneven print across	Uneven tension between gravure and transfer rollers
Ink build-up on drop chute	Ink drying too slowly (thin it with approved solvent)
Poor adhesion of ink to tablet	Wrong ink is being used
	Excessive amounts of lubricant used

10 Sorting and Packing

Sorting can be accomplished either manually through 100% visual inspections or by mechanical means. The mechanical sorters are culling devices used to remove tablets with dimensions outside of thickness and/or diameter specifications. Most products are usually sorted and then passed through a metal detector prior to being packed for shipping. Although sorters are set up specifically for a certain size product, its performance is not 100% effective. Sorters can damage tablets (scuffing, breaking apart twins, chipping coatings, etc.)

There are different types of sorters made by different companies. Thickness-only sorters: e.g., a Pro-Quip 3-Lane sorter and Bohle 3-Lane sorter. These thickness-only sorters are most often used between membrane coating and drilling to remove twins. In addition, there are sorters capable of both thickness and diameter sorting: e.g., Bohle 5-Lane and Eriez sorters. The Bohle 5-Lane and Eriez sorters, which also check for proper diameter, are primarily used between printing and packing. Most of the 100% visual inspections are accomplished on Lakso tables.

Sorting is accomplished between membrane coating and drilling; if the lot fails the membrane coating AQL, then sorting will remove the defects. These failures are almost always for twinning, and sorters do a reasonable job of culling the twins from the lot. A Pro-Quip and Bohle 3-Lane thickness sorters may be used primarily to sort after membrane coating AQL failures.

Sorting can be done concurrently with laser drilling per MF. However, the sorters are much slower than the laser drills. Most often the first sub lot will be sorted prior to drilling, and the remaining sublots sorted concurrently with drilling. Separation of sorted and unsorted cores and labeling is critical, especially during concurrent operations. The sorters have controlled vibrating feed hoppers that meter tablets to the sorting mechanisms. For example, the Pro-Quip thickness sorter uses three sets of sloped counter-rotating rollers to cull over- and under-thickness tablets. The distance between the rollers is adjustable for different products and is varied from end-to-end using two adjustment knobs. During set-up the operators use pin gauges, pieces of specific-diameter steel rods, to set the rollers. The pin gauges should just fit between the rollers at points right over the top of the V-shaped diversion bars that are just below the rollers. The under-sized tablets will drop through prior to the first diversion bar, ending up in a drug waste bag for under-sized tablets. The over-thickness tablets will continue past the second diversion bar and drop into a bag for over-sized tablets. The tablets that are acceptable should fall through the rollers between the two V-shaped diversion bars and end up in the in-process bag. They slowly move down the rollers because of the slope.

During set-up, the operators adjust the distance between the discs and the position of the oversized and acceptable tablet guides so that tablets are directed into the appropriate tray or chute. To make the adjustments, the operators use standard tablets (half at the minimum thickness and half at the maximum thickness

DOI: 10.1201/9781003224716-10

specification). One side of each pair of discs is simultaneously adjusted via a knob on the end of the common shaft that runs through them. Between this adjustment and the guide placement, only acceptable thickness tablets should end up in the in-process bag. Oversized tablets can be reprocessed once, since tablets, if fed too rapidly, can clog the discs and end up in the oversized tray. Under-thickness tablets are never reprocessed.

Mechanical sorters (e.g., Eriez models) use vibrating plates and racks to cull over- and under-sized tablets. The tablets are initially fed from the hopper onto a stainless-steel V-shaped parallel bars that make up the grizzly rack that keeps over-diameter tablets from dropping through. The rest of the tablets drop onto a set of perforated stainless-steel plates. These plates, like the rotating diameter sorting discs, are drilled with specific size holes. The top plate prevents over-diameter tablets from dropping, and the lower plate allows under-diameter tablets to drop through, the good tablets end up between the plates. The good tablets are then fed to another grizzly rack that cull for under-thickness that allows under-thickness tablets to drop through. The good tablets are then fed into an in-process bag.

If a 100% inspection is required, it is most often accomplished on a Lakso table. The tablets are delivered to the table via a vibration-modulated feed hopper. The table has a conveyor belt on an upper level that initially moves the tablets past the first operator, who, while sitting at the side of the belt, inspects the first side of the tablet. The tablets are then turned over as they leave.

The final step in the manufacturing process is to run the tablet through a metal detector prior to packing. Metal detection is usually accomplished by catching the tablets as they come out of the sorter on a conveyor belt that passes them through the metal detector throughput chute. If metal particles, above a specified size, are magnetically detected, a reject gate diverts the tablet(s) into a drug waste container. Prior to start-up, the metal detection sensitivity level is checked by passing ten test sample tablets through the detector. All test samples must be rejected. The test samples are selected based on product. Low iron products, those without iron oxide coloring, are checked using test samples with 0.5 mm stainless-steel spheres encased in Lexan. High iron products, those containing iron oxide, use test samples containing 1.0 mm spheres.

After completion of each drum, the operators recheck the sensitivity levels to verify the metal detector operation by passing a single test sample through the detector. If the recheck fails, the tablets in the last drum are passed through the detector again.

CRITICAL ELEMENTS OF THE SORTING PROCESS

1. Proper set-up of sorter using pin gauges or standard tablets.
2. Monitoring the sorter during the process to prevent scuffing and of the sorting mechanisms.
3. Use the correct test sample tablets to set up the metal detectors.
4. Use metal detector to check tablets from sorter. Document metal detection tests for each shipping drum.

PACKING OVERVIEW

Following printing/sorting, the tablets are prepared for shipment to clients in bulk. The packing requirements are product specific. SOP contains specific packing instructions for products that do not have packing instructions included in the MF. MF instructions take precedence over those in the SOP. Tablets are normally packed in fiber or steel drums inside two independently sealed poly bags. Strips of bubble pack are placed across the bottom and up the sides and then folded across the tops of the tablet bags in the drums. The excess head space is also filled with bubble pack to prevent the product from shaking within the drum. Desiccant units to absorb moisture are taped securely to either the inner or outer bag per instructions. The bags and drums are closed using tamper-evident seals.

Various labels are attached to the internal poly bags and to the exterior of the drum. The shipper labels are printed and issued specifically for a lot and contain detailed information concerning product, lot #, manufacturer, client, and storage. They need to be checked closely against the samples in the MF or accompanying specification sheet. The labels show a manufacturing date that is based on the start or finish of a specific production step; it varies by product and is specified in the MF. Process operators annotate the shipper labels with tare, net, and gross weights. They also annotate the drum number and the total number of drums in the lot. No cross-outs/corrections are allowed on the labels. Unused or damaged labels are saved for reconciliation. The final step is to waterproof the labels using clear 2″ tape placed horizontally across the label from bottom to top, overlapping each piece of tape.

11 Capsule Filling

Capsule filling is a complex process whereby the encapsulated product must be well developed with an acceptable flow and a reproducible batch-to-batch density to ensure mass uniformity. Key considerations in capsule filling, in principle, any type of formulation, may be dosed into hard capsules, from blends or granules, to coated pellets, to other oral dosage forms such as small tablets, micro-tablets, smaller capsules, microspheres, including various combinations of mentioned forms. It is also possible to fill liquids, provided that the material of the capsule shell (generally gelatin, or other alternatives) is not soluble in the solvent used in the formulation.

Due to the need for a plasticizer in the capsule shell formulation, conventional hard gelatin capsules have a high-water content (13%–16%). For this reason, hygroscopic products can absorb moisture from the capsule, causing the capsules to become brittle, which can break under mechanical stress. In addition, such transfer of moisture to the contents of the capsule could generate problems of physical stability (crystalline form) or chemical stability of the API. Capsules with a low moisture content can be developed, either by using plasticizers other than water or other polymers, such as hydroxypropyl methylcellulose (HPMC), which is a commonly used alternative. Mixtures with materials containing reactive aldehydes are not suitable for capsule filling, as cross-linking of gelatin is probable, reducing the capsule's solubility. It is possible to fill powders, granules, non-aqueous liquids, non-aqueous gels, and thermo-setting formulations into capsules. For solids, powders with poor flow properties can be problematic because of challenging fill weight control. Liquids with very low viscosity can leak from two-piece shells soon after filling.

The choice of capsule size and fill weight is dictated by the unit dose requirements and the used formulation. Once the formula of the contents of the capsule is developed and the dose weight is defined, the most appropriate capsule size is determined. Each size of capsule has a defined volume, and by knowing the density of the fill mixture, the volume and weight in each capsule are established. Mixtures (powder or granulate), knowing the tapped density makes it possible to determine the most suitable capsule size. For pellets or micro-tablets, where there is no possibility of compaction, it is necessary to use bulk density instead. All capsule suppliers provide capsule size tables that facilitate the choice of capsule size.

Capsule filling provides great potential for clinical trials, whether it's filling just the API for first preclinical or clinical trials to filling other dosage forms in capsules (tablets or smaller-sized capsules) for blind or double-blind clinical trials. Capsules intended for clinical trials often have different geometries or dimensions from the capsules used in production and, therefore, might require specific machine formats. Typically, the batch sizes for clinical trials are smaller than for commercial production, so in many cases, it is possible to fill capsules manually.

For small numbers of capsules for clinical studies, it may be feasible to hand-fill capsules, either with neat drug or with a simple powder blend, or use a precision

DOI: 10.1201/9781003224716-11

powder dispenser, such as the Xcelodose (Capsugel). Another approach for relatively small numbers of capsules is to flush-fill, which involves holding an array of opened capsule bodies in a Perspex frame and adding, spreading out the drug, or formulation across the capsule shells to fill them all to the brim, followed by adding the caps and closing the capsules. Difficult to handle powders (e.g., micronized drugs, poor flow powders) can be filled into capsules either by hand-filling or by flush-filling (although due consideration should be given to the quantities of capsules required).

For larger clinical trials, automation becomes important. For a batch size greater than a few thousand capsules, an automated filling approach is required, which demands the use of a suitable, free-flowing powder blend or granular formulation.

For commercial production of capsule products, speed and ease of manufacture is important, and therefore, automated capsule filling machines are used. There are two types of automated capsule-filling machines: the dosator type and the dosing disc type. For liquid formulations, very low viscosity liquids should be avoided due to the potential for leakage. For powder formulations, good flow properties are important to ensure satisfactory fill weight control.

The main challenge when filling liquids in hard gelatin capsules is to find the right solvent (i.e., one that does not interact with the capsule material). From a technological standpoint, the capsule filling machine requires a specific station to fill liquids, and the capsules must undergo an additional subsequent band-sealing process to prevent leakage of the contents of the capsule.

As far as powder filling is concerned, the main challenge is to achieve a formulation with good flow properties, which guarantees the correct filling of the dosing systems, and therefore, a good uniformity of mass of the contents of the capsule.

Unlike an automatic filling machine, where there is a weight control transfer of the product to the capsule, the manual filling of powder is done volumetrically. This implies that the whole body of the capsule must be filled. Changes in API or final blend density present challenges to fill the exact amount of blend. In cases where the density is lower, it is possible to use manual compaction systems, but they are less effective and introduce a high variability to the contents of the capsule.

The main advantage of manual encapsulation is that the investments in equipment and formats are much lower and that the times dedicated to associated activities, such as adjustment and cleaning, are drastically reduced. Another advantage of manual filling is that there is no dead volume, which makes it possible to encapsulate all the powder, something that is useful when manufacturing very small batches for clinical trials or conducting formulation tests.

Speed, reliability, and the ability to adjust the dose are the main advantages of automatic machines. Its main disadvantages include the need to invest in expensive equipment, the longer time needed to perform adjustments to the batch, format changes and cleaning, and the need to manufacture bigger batch sizes compared with manual filling.

When powder filling, the main challenge is poor capsule fill weight control. When filling liquid or semi-solid formulations, splashing of the liquid during the capsule filling can occur, making it necessary to reduce machine output speed or increase time for pump stroke. In addition, tailing during filling stroke for gels and semisolids can lead to variable fill weights and contamination of capsules, making it necessary to adjust the speed of pump stroke and/or increase the temperature of the formulation during filling.

An automatic encapsulating machine is a complex equipment. Each capsule format is composed of numerous pieces, which must be perfectly aligned and adjusted to avoid the opening or breaking of capsules. When capsules are broken, they can release their contents, staining the entire batch produced. Finally, hard gelatin capsules eventually undergo a cross-linking reaction that renders the gelatin less soluble and may affect the release of the capsule contents.

Capsule-Filling Machine Parts

Alignment Plate – This allows to quickly load the capsule bodies and capsule caps into the slots of the plates without having to load them one by one.

Body Plate – This holds the longer side of the capsule aka "The Body."

Cap Plate – This holds the shorter side of the capsule aka "The Cap."

Middle Plate – This comes into play when pressing the capsules together. Once you press the capsules together, the middle plate holds the finished capsules to make it easier to lift them out of the machine all at once.

Tamper (Powder-Pressing Plate) – This packs the powder down into the capsules once the powder is loaded into the capsule bodies. Typically, use the tamper 2–3 times to make sure caps get fully packed. Lighter/fluffier powders require more tamper runs than heavier/denser powders.

Spatula (Powder Spreader) – This makes it easier to load the powder into capsules. Use after dumping the powder above the capsule bodies to evenly disperse powder into the capsule bodies.

Spill Guard – This goes around the edges of the body plate when loading capsules to prevent the powder from spilling over the edges of the plate. When done loading capsules, the spill guard also makes it easier to dump out the excess powder without making a mess.

To find out how many milligrams are in each capsule, a pocket milligram scale is used to calculate the density of powder.

A soft gel is an oral dosage form for medicine like capsules. They consist of a gelatin-based shell surrounding a liquid fill. Encapsulators, also called capsule fillers, capsule filling machines, or encapsulation machines, are mechanical devices commonly used for industrial and pharmaceutical purposes. These machines are used to fill empty soft or hard gelatin capsules of various sizes with powders, granules, semi-solids, or liquids substances containing active pharmaceutical ingredients or a mixture of active drug substances and excipients. This process of filling empty capsules with substances is termed encapsulation.

Capsule-filling machines all have the following operating principles in common

- Rectification (orientation of the bad gelatin capsules).
- Separation of capsule caps from bodies.
- Dosing of fill material/formulation (filling the bodies).
- Rejoining of caps and bodies
- Ejection of filled capsules.

Various types of encapsulation machines are available; these machines are selected based on the requirement of the manufacturer/nature of the capsule (hard capsule or soft capsule), and the quantity of capsule to be manufactured.

Encapsulators used in the encapsulation of hard gelatin capsules can be classified or said to be of three types:

- Manual/hand-operated capsule-filling machine
- Semi-automatic capsule-filling machine
- Automatic capsule-filling machine

These types of encapsulators consist of

- A bed of about 200–300 holes
- A loading tray with about 200–300 holes
- A powder tray
- A pin plate with about 200–300 pins
- A sealing plate with a rubber cap
- A lever
- A cam handle with a loading tray of about 250 holes on the average

A hand-operated capsule-filling machine can produce about 6,250 capsules per hour. This machine is used by small-scale manufacturers and hospitals for extempore preparations.

Semi-automatic encapsulators (semi-automatic capsule-filling machines) combine both manual and automatic methods of capsule filling, thus can be said to be partially automated. Its operation is simple, and the equipment meets the hygiene requirements for its use in the pharmaceutical industry. Its simple design and robust construction (which ensures long life and trouble-free operation) as well as the use of stainless steel and non-corrosive approved materials in the construction of contact parts (which eliminates contamination and facilitates easy cleaning after use) make the machine suitable for filling powders and granular materials in the pharmaceutical and health food industries. Sandwich of cap and body rings are positioned under rectifier to receive the empty capsule, and the caps are separated from the body by pulling vacuum from beneath the sings. The body rings are then positioned under the foot of the powder hopper for the filling process. The cap and body rings are rejoined and positioned in front of pins which push the bodies to engage pins which push filled bodies. The plate is then swung aside, and the pins are used to eject the closed capsule.

Automatic encapsulators are capsule-filling machines that are developed and designed to automatically fill empty hard gelatin capsules with powders and granules. They are used in the large-scale production of capsules. Automatic capsule-filling machines are extremely durable and reliable when it comes to capsule filling and maintenance of the integrity of the filled capsules. Automatic encapsulators can also work as a complete system of fully automatic capsule-filling line by attaching additional equipment as an online capsule-polishing machine, dust extractor, damage capsule sorter, and empty capsule ejector.

TABLE 11.1
Capsule Sizing

Density (g/mL)	0 Capsule Size (mg)	00 Capsule Size (mg)	000 Capsule Size (mg)
0.6	400	575	820
0.8	540	755	1,095
1.0	675	955	1,375
1.2	815	1,135	1,640

Capsule Size: 0#, 00#, 000#,1#,2#,3#,4#,5# are available.

Microparticles are particles with dimensions between 1×10^{-7} and 1×10^{-4}m. The lower limit between micro- and nano-sizing is still a matter of design criteria. To be consistent with the prefix "micro" and the range imposed by the definition, dimensions of microparticles should be expressed in µm. However, general acceptance considers particles smaller than 100nm nanoparticles. Anything larger than 0.5µm and anything smaller than 0.5mm is considered microparticles. Very often particles with dimensions more than 100nm are still called nanoparticles. The upper range may be between 300 and 700nm, so this would give a size definition for microparticles of 0.3–300 or 0.7–700µm.

Microspheres are spherical microparticles and are used where consistent and predictable particle surface area is important. In biological systems, a microparticle is synonymous with a micro-vesicle a type of extracellular vesicle. In addition, hollow microspheres loaded with drug in their outer polymer shell were prepared by a novel emulsion solvent diffusion method and spray drying technique. Drugs can be formulated as hydrodynamically balanced system (HBS) floating microspheres. The following are drugs which can be formulated as microsphere: Repaglinide, Cimetidine, Rosiglitazone, Nitrendipine, Acyclovir, Ranitidine HCl, Misoprostol, Metformin, Aceclofenac, Diltiazem, L-Dopa, beneseragide, and Fluorouracil.

An important implication of molecular encapsulation whereby a molecule is prevented from contacting other molecules that it might otherwise react with. Thus, the encapsulated molecule behaves very differently from the way it would when in solution. The encapsulated molecule tends to be extremely unreactive and often has much different spectroscopic signatures. Compounds that are normally highly unstable in solution, such as arynes or cycloheptatetraene, have been successfully isolated at room temperature when molecularly encapsulated (Table 11.1).

12 Safe Handling of APIs and Drugs

The purpose of this guideline is to describe procedures to safely handle Active Pharmaceutical Ingredients (APIs) and drug products at a company. Company places a strong emphasis on protecting its employees, which are the most valuable resource, and respecting the environment in which they live. It is usually a company policy to safeguard the environment and the health and safety of its employees and to comply with all applicable regulations. Company produces and handles materials that may not be specifically covered by regulations in all countries that it operates. Company is committed to protecting the health of its employees regardless of hazards being regulated or not. Company develops these guidelines and usually continually evaluates risk reduction mechanisms for safely handling materials which may otherwise create a hazard. These guidelines are a global standard for a company, and they are implemented according to local site-specific considerations. If a conflict arises between local laws and regulations and these guidelines, safety, hygiene, and health considerations will be a priority but compliance with all local regulations and laws will always be required.

These guidelines apply to all sites which handle APIs and drugs in all dosage forms and all processing stages and activities including but not limited to API synthesis, parenteral preparations, oral solid dose formulation and packaging operations, validation, laboratories, research and development, cleaning, waste disposal, and maintenance. These guidelines do not apply to solvents, cleaning agents, laboratory reagents, and other chemicals for which health and safety data is provided in the supplier's material safety data sheets (MSDSs).

These guidelines do not apply to biological organisms, biological preparations, vaccines, or fermentation activities. Separate containment handling requirements are necessary for these situations since the risk is based on microorganism biosafety level as opposed to the Occupational Health Category (OHC). Handling highly allergenic materials (e.g., penicillin and cephalosporin), cytotoxic, cytostatics, hormones and analogs, biologics, or other related highly active/potent drugs, should be conducted according to a company Global Quality Guidelines such as manufacturing control of beta-lactam antibiotics, cytotoxic compounds, hormones, biologics materials, and potent products.

Establishing the OHC or in some cases the Occupational Exposure Limit (OEL) is an important responsibility that is required in authoring MSDSs for APIs manufactured by company. The responsibilities cover overseeing the development of control strategies, regulatory updates, and guidelines, and updating OHC and exposure controls as required. In addition, monitoring the effectiveness of the implementation of these guidelines at company facilities. Providing technical guidance and support to company sites and corporate departments regarding hazard assessment, risk

DOI: 10.1201/9781003224716-12

assessment, exposure controls, containment, industrial hygiene monitoring, and medical surveillance is also in scope. Major tasks include evaluating available containment and exposure control technologies and updating the Control Matrix Guidelines and Toolbox, developing and implementing technical training programs in order to implement the guidelines, and monitoring the effectiveness of the implementation of these Guidelines at company sites.

Supporting R&D and production mangers regarding safe handling procedures for APIs and drugs to ensuring that all APIs are assigned an appropriate OHC prior to handling them is required. Following up on the implementation of corrective actions is required as a result of industrial hygiene (IH) audits, sampling reports, gap assessments, and risk assessments. API implementations require working with engineering, manufacturing, R&D, corporate EHS, or other affected departments to conduct risk assessments in order to provide input and appropriate recommendations and evaluating available containment and exposure control technologies by comparing them to the appropriate control target.

Other activities in developing training programs for company employees with EHS support and developing and implementing procedures for handling situations when employees and/or local EHS professionals report any concerns or problems with the employees' safety, health, and reproductive concerns by managing the medical surveillance logistics and providing logistical support for medical surveillance.

Operations site manager/R&D manager, or their delegates, at each facility are responsible for when introducing any new API, new process, or new equipment. R&D manager is responsible to provide EHS with updated information to allow reevaluation of the OHC. R&D manager is responsible to notify EHS when a new product gets FDA marketing approval.

Ensuring that hazard and risk assessment are conducted before work begins with a new product or process, and the appropriate controls are implemented, EHS is provided with information necessary for risk assessment of the new APIs, equipment, or processes (e.g., process flows and equipment lists). Implementation of these guidelines for new facilities, processes, equipment, and implementation of the gap analysis plan for existing facilities, processes, and equipment are all important considerations.

DEFINITIONS

Active Pharmaceutical Ingredient (API) – Any chemical or biochemical substance having pharmacological activity that may be used in the treatment or prevention of an illness or injury.

Control Matrix – A table that correlates the engineering containment solutions or the facility requirement to the OHC for which it is intended to be used.

Control Target for Design – The reference value of airborne concentration of a given API for purposes of facility and unit operation design to be achieved through containment and engineering controls. The reference value of the control target for design is the lower limit of the OHC or the OEL (when established).

Exposure Control Target – The reference value of airborne concentration of a given API below which the exposure is acceptable. The reference value of the exposure control target is the geometric mean of the relevant OHC or the OEL (in case an OEL has been established).

Containment Strategy – A series of decisions designed to contain materials assigned to a specific OHC. Containment strategies should be developed in accordance with the hierarchy of controls.

Developmental Toxicity (Teratogenicity) – The ability of a chemical or other agent to adversely affect the normal development of the fetus and/or the newborn.

Potent Compound – For the purposes of these guidelines, a potent compound is a compound (API) with assigned OHC of 4, 5, and 6.

HIERARCHY OF CONTROLS

Elimination of hazard (substitution, tech transfer, or appropriate site transfer).

Control at the source – primary control (closed system, containment solutions).

Control of the environment (secondary controls such as ventilation, HVAC, designated work areas).

ADMINISTRATIVE CONTROLS

Personal Protective Equipment (PPE)

Industrial Hygiene (IH) Monitoring – The quantitative measurement of exposure to hazards in the air or on surfaces in order to determine the extent of exposure to workers and ensuring exposure controls are employed that adequately protect workers from the exposures.

Material Safety Data Sheet (MSDS) – A document prepared by a chemical manufacturer describing hazardous properties of a chemical and appropriate ways to handle and dispose of it.

Occupational Exposure Limits (OEL) – Maximum exposure limit level of material in the air considered to be acceptable for healthy employees and without any negative health effects. Limits are usually expressed as 8-hour time-weighted averages (TWA) (weight per unit volume, e.g., μg/mL) for exposures of 40 hours a week over a working lifetime.

Occupational Health Category (OHC) – A set of consecutive airborne concentration ranges ("bands") used to classify the maximum permitted exposure levels for a range of APIs, based on their health hazards and toxicological properties and potency, which are used to communicate hazard potential and control measures required for handling them safely. The OHCs ranges, design conditions, and operating parameters are shown in Tables 12.1–12.9.

HAZARD ASSESSMENT

Hazard assessment is the first stage of a risk assessment process. It is used to define and communicate hazard potential of the chemical entities used in the process. Prior to handling, all APIs must be assigned to an appropriate OHC. Local EHS shall notify company corporate EHS of any existing products that are already being

TABLE 12.1
Occupational Health Category (OHC) against Occupational Exposure Limit (OEL)

OHC	OEL [μg/m³]
1,2	$100 \leq X < 3,000$
3	$10 \leq X < 100$
4	$1 \leq X < 10$
5	$0.1 \leq X < 1$
6	$X < 0.1$

TABLE 12.2
Occupational Health Category (OHC)

Air Changes/ Hour (AC/h)	OHC 4	OHC 5	OHC 6	OHC 1/2	OHC 3
Air recirculation and % of fresh air (% FA)	($1 \leq X < 10$ μg/m³)	($01 \leq X < 1$ μg/m³)	($x < 0.1$ μg/m³)	($100 \leq X < 3,000$ μg/m³)	($10 \leq X < 100$ μg/m³)

TABLE 12.3
Facility/Infrastructure General Considerations

Designated Area with Not Required	Designated Area with Restricted Access Is Required
A negative differential air pressure relative to surrounding areas is not required	A negative differential air pressure relative to surrounding areas is required (10 Pa minimum; 15 Pa typical). A separate HVAC system for OHC > 3. Monitoring of the air pressurization system and failure alert are required. Unless in controlled containment: Pressure gradients are not required
6–12 AC/h typical	15–20 AC/h typical
100% FA (100% once-through) or 20% fresh air with recirculation if possible	Unless in controlled containment: 6–10 AC/h typical

handled but have not had an OEL or OHC developed and documented by company. The R&D project manager should ensure that hazard and risk assessment are conducted before work begins with a new product or process, and appropriate controls are implemented.

New Chemical Entities (NCE) are assigned an OHC by innovative R&D or Innovative department. Innovative R&D is responsible for establishing and communicating the OHC to the Global EHS department. In instances where developing an OHC may not be feasible for a period such as in early stages of new drug development, the default shall be an OHC 4. In instances where the API is in the pre-approval

TABLE 12.4
Occupational Health Category (OHC) – Design Conditions

Condition	OHC 1, 2, 3	OHC 4, 5, 6
HEPA filter on	Terminal HEPA filters – not required	Terminal HEPA filters – required
HEPA filter on	HEPA filters not required	Room FEPA filters – required
Filter change	Regular	Bag in/bag out (BIBO – required
Personnel air lock	Not required	Required + mist shower + sink for cleaning
Separate gowning	Not required	Required + mist shower + sink for cleaning
Material air lock	Not required	Not required
Maintenance	Special attention to maintenance	Special attention to maintenance
Spills	Spill containment. Personnel protection	Cleaning and decontamination – required
Dust collectors	BIBO HEPA filters on exhaust	BIBO HEPA filters on exhaust
PPE	Respiratory protection	Full protection/Tyvek suit/double gloves
Waste disposal	Collect and treat	Controlled containment/ respiratory Protection
Equipment/material	Clean contaminated item	Complete decontamination

TABLE 12.5
Occupational Health Category (OHC) – Design Conditions Designations

	OHC ½ ($100 \leq X < 3{,}000$ μg/m³)	OHC 3 ($10 \leq X < 100$ μg/m³)	OHC 4 ($1 \leq X < 10$ μg/m³)	OHC 5 ($0.1 \leq X < 1$ μg/m³)	OHC 6 ($X < 0.1$ μg/m³)
Designated area with access control	Not required		Designated area with restricted access is required (one access control for a group of suites is acceptable)		
Pressure gradients	A negative differential air pressure relative to surrounding areas is recommended. (10 Pa minimum; 15 Pa typical)		A negative differential air pressure relative to surrounding areas is required. (10 Pa minimum; 15 Pa typical.) A separate HVAC system for OHC > 3. Monitoring of the air pressurization system and failure alert are required		
Air changes per hour (AC/h)	15 AC/h typical		15–20 AC/h typical. Unless in controlled containment: 6–10 AC/h typical.		
Air recirculation and % of fresh air (% FA)	100% FA (100% once-through) or 20% fresh air with recirculation if possible. Note: If solvents are used, 100% FA is required		100% FA (100% once-through) or 20% fresh air with recirculation if possible. Note: If solvents are used, 100% FA is required		

phase and there is a limited amount of data, it is imperative to revise the OHC as more data becomes available. Intermediates that are structurally like the API (based on R&D evaluation) will be considered to have the same OHC as their API. This will be the default choice unless available information justifies a different OHC.

TABLE 12.6
Occupational Health Category (OHC) – Design Parameters

Condition	OHC 1, 2, 3	OHC 4, 5, 6
HEPA filters on – supply air	HEPA filtration of supply air required. Terminal HEPA filters are required for recirculated air	Terminal HEPA required
HEPA filters on – return air	HEPA filters required	HEPA filters in room, roof, and controlled containment
Filter change	Regular	BIBO (Bag In/Bag Out) method
Personnel air lock	Required. Recommended for controlled containment	Required. Required for controlled containment
Separate gowning/ degowning areas	Required for new facilities	Required + mist shower + sink for cleaning
Material air lock	Required + material decontamination area	Required + material decontamination area
Maintenance/spills	Personnel protection. Spills to be cleaned	Personnel full protection. Cleaning/ decontamination under full protection: respirator, Tyvek suit, double gloves
Dust collectors	No recirculation. BIBO HEPA filters on exhaust	BIBO on exhaust. No recirculation
PPE	Full protection air respirator, Tyvek suit, double gloves	Full protection: respirator, Tyvek suit, double gloves
Waste disposal	Collect and treat	Collect, contain, label, and treat

An OEL will be established in specific cases by the Global EHS division for the following reasons:

1. Regulatory requirements.
2. Special cases of high-risk APIs.
3. Special design requirement (multipurpose suite).

Once an OHC has been assigned, an appropriate containment and control strategy for the category will be implemented. OHCs will be reevaluated and updated periodically or when new information becomes available.

RISK ASSESSMENT

Risk assessment is the process by which a qualitative and/or quantitative probability and severity of a potential exposure to a given hazard is evaluated. The risk assessment process uses a certain amount of professional judgment and should be conducted by EHS, engineering, production, and other affected departments. The risk assessment is used to determine which appropriate process and facility exposure controls/containment, administrative controls, and PPE are necessary in order to safely handle the API.

TABLE 12.7
Occupational Health Category (OHC) – Design Conditions – Powders

	OHC 1/2 ($100 \leq X < 3{,}000$ μg/m³)	OHC 3 ($10 \leq X < 100$ μg/m³)	OHC 4 ($1 \leq X < 10$ μg/m³)	OHC 5 ($0.1 \leq X < 1$ μg/m³)	OHC 6 ($X < 0.1$ μg/m³)
Containment	**Control Strategy**	**Engineering Matrix**			
Dispensing/weighing – wet powders	–	LEV	LEV or down flow booth	Down flow booth with barriers or curtains, contained transfer ports or isolators	Isolator with contained transfer ports
Dispensing/weighing – dry powders	LEV	LEV or down flow booth	Down flow booth with barriers or curtains, contained transfer ports or isolators	Isolator with contained transfer ports	
Charging/feeding – wet powders	LEV		Ventilated enclosure, down flow booth, direct connection between processing units or isolator	Appropriate contained transfer technology, direct connection between processing units or isolator	Direct connection between processing units or isolator with contained transfer ports
Charging/feeding – dry powders	LEV	LEV and a flexible direct connection	Appropriate contained transfer technology, (e.g., PTS, DCS, etc.), direct connection between processing units or isolator	Direct connection between processing units or isolator with contained transfer ports	

TABLE 12.8
Occupational Health Category (OHC) – Design Conditions - Powders Operations

	OHC 1/2 ($100 \leq X < 3{,}000$ µg m³)	OHC 3 ($10 \leq X < 100$ µg m³)	OHC 4 ($1 \leq X < 10$ µg m³)	OHC 5 ($0.1 \leq X < 1$ µg m³)	OHC 6 ($X < 0.1$ µg m³)
Seeding – dry powders		-	Appropriate contained transfer technology (e.g. SBV)		Isolator with contained transfer ports
Powder sampling (when dedicated sampling port is not available)	LEV	LEV or down flow booth	Down flow booth	Isolator with contained transfer ports	
TD centrifuge discharge		Not suitable			
BD centrifuge discharge	Suitable enclosure (e.g., inflatable seal)		Suitable continuous liner	Isolator with contained transfer ports	
Horizontal centrifuge discharge	Suitable enclosure (e.g., inflatable seal)		Suitable continuous liner or other appropriate contained transfer technology	Isolator with contained transfer ports	
Inverted basket centrifuge discharge	Suitable enclosure (e.g., inflatable seal)		Suitable continuous liner or other appropriate contained transfer technology	Isolator with appropriate contained transfer ports	
Other liquid filtration when solids are waste		Thoroughly wash out the unit before opening for disposal	Safely dispose after proper wetting. Use disposable bags/cartridges if possible. Dispose in contained manner, e.g., while using "Bag in Bag" or "Cartridge in Bag" technology		
Filter dryer discharge	LEV or suitable enclosure	Suitable enclosure (e.g., inflatable seal)	Suitable continuous liner	Isolator with contained transfer ports	
Fluidized-bed dryer	Direct connection or bowl inverter and LEV and WIP/CIP recommended		Contained transfer technology and WIP or CIP or isolator	Not recommended (change drying technology)	
Tray dryer	LEV or down flow booth		Down flow booth with barriers or curtains, or isolator, or change drying technology	Use isolator for loading and unloading or change drying technology	
Lyophilizing/freeze drying	LEV		Down flow booth with barriers or curtains	Isolator for vials: wash down system	
Vacuum dryers charge/discharge	Direct connection		Direct connection with appropriate contained transfer device and WIP/CIP		

TABLE 12.9
Occupational Health Category (OHC) – Design Conditions – Processes

	OHC 1/2 ($100 \leq X <$ 3,000 µg m^3)	OHC 3 ($10 \leq X < 100$ µg m^3)	I	OHC 4 ($1 \leq X < 10$ µg m^3)	OHC 5 ($0.1 \leq X < 1$ µg m^3)	OHC 6 ($X < 0.1$ µg m^3)
Blending (V, Bin, cone, etc.)	LEV	ELV and direct connection	Appropriate contained transfer device and WIP/CIP			
High sheer mixing	LEV	LEV and direct connection	Appropriate contained transfer device and WIP/CIP; isolator for product heel removal			
Coating	LEV		Appropriate contained transfer device and WIP/CIP			
Milling	LEV and direct connections or down flow booth		Direct connection and appropriate contained transfer device and WIP/CIP or isolator			

The qualitative risk assessment shall include, but is not limited to, the following:

1. Identification of APIs involved and their OHC/OEL.
2. Identification of potential routes of exposure.
3. Recognition of process and activity-dependent factors and any other factor in the area that may lead to exposure. Among factors to be considered are:
    a. The amount and percentage of API handled.
    b. Physical form (powder, liquid, etc.) and properties of the API.
    c. The number of transfers.
    d. The number of batches per year.
    e. Equipment type.
    f. Process duration and frequency.
    g. Facility and process controls/containment.
    h. Dustiness of the material.

A quantitative assessment, such as industrial hygiene (IH) monitoring, will be considered for the determination of:

1. Effectiveness of exposure control.
2. Employee's exposure assessment.
3. IH monitoring will be carried out.

The IH monitoring is designed to obtain representative exposure results including timing, duration, air volume, and number of samples. IH data are compared to the appropriate exposure limits such as the geometric mean of the OHC, OEL, etc. Professional judgment by EHS may be required during the outcomes and interpretation of the risk assessment process to determine how best to achieve the desired risk reduction and reevaluation for a circumstance and steps.

Control Strategy

The strategy aims at achieving adequate worker's protection and reduction of exposure and risk to an acceptable level. The emphasis should be on the implementation of the hierarchy of controls with engineering control (primary controls) being the highest priority. The Control Design Target should be the lower limit of the OHC or the OEL (when established). The Exposure Control Target should not exceed the geometric mean of the relevant OHC or the GEL (when established).

A "Controlled Containment System" exists when: The results of IH monitoring are below the OEL or the geometric mean of the relevant OHC. Adequate certainty is demonstrated by the exposure assessment (e.g., statistical evaluation, number of samples taken, and repetition of studies). The operational reliability of the system is adequate. There are no other sources of exposure in the area. Where "Controlled Containment System" exists, secondary control such as facility controls, administrative, and PPE may be reduced. This is subject to an approval of company EHS professional(s). When given judgment disagrees with the matrix recommendations, the rationale of the chosen control strategy should be documented and approved by EHS in advance.

Where an exposure assessment indicates levels above the geometric mean of the relevant OHC, or above the OEL, additional protection measures (PPE respiratory protection) are required until controls are implemented and verified to reduce exposure below the target levels. A combination of engineering and administrative control measures as well as PPE is acceptable for OHC 1, 2, and 3 only. Administrative control and PPE may be used as interim control measures for OHC 4, 5, and 6 when:

i. target is not met by the already established engineering control, and additional protection is needed as a temporary solution until the full solution is applied.
ii. Emergency protection is required.
iii. The main tool for the implementation of the "Control Strategy" is the Control Matrix Guidelines.

The Control Matrix Guidelines is a tool which is based on:

i. Accumulated knowledge and experience in the pharmaceutical industry.
ii. IH monitoring data.
iii. The Control Matrix Guidelines is updated periodically.
iv. The Control Matrix Guidelines is intended as a suggested approach that has been demonstrated to be effective under similar circumstances.
v. Each application of the guidelines should be assessed on a case-by-case basis, based on cumulative experience and professional judgment both of which should be used to decide upon the control alternatives for specific operation within a given range or ranges of OHC

Once the controls are implemented, EHS performs IH monitoring to verify their effectiveness and to ensure that they reduce employees' exposure below exposure targets with adequate certainty.

New facility, processes, or a new API:

Controls for new facilities, new processes, or a new API should be constructed according to the strategy presented in these guidelines as part of the conceptual design. The design should follow the Control Matrix Guidelines. The design will be captured in the User Requirements Specification (URS) and Basis of Design (BOD). The design will be approved by the relevant departments including EHS professionals.

A risk assessment for existing facilities and processes must be performed to evaluate the current degree of compliance with the Control Matrix Guidelines. Existing facilities or processes shall undergo risk assessment to cover the gap analysis between existing infrastructure conditions and guidelines requirements.

Introduction of new or modified equipment, techniques, work procedures:

According to the results of the risk assessment, an action plan will be established. The action plan will address any issues identified in the risk assessment, including timelines for completion and responsible parties. Containment projects should be prioritized according to the risk of potential employee exposure, number of employees involved, batch size, frequency of synthesis or manufacture, etc.

When a process/product is being transferred from R&D or from one site to another, the hazard and risk assessment information should be included with the technical transfer package. This package should include but not be limited to the following information:

i. The OHC
ii. Special considerations such as cytotoxicity, reproductive concerns, sensitizers, etc.
iii. IH monitoring information
iv. MSDS
v. Description of appropriate control measures (engineering controls, PPE, etc.) which were applied in the R&D or former manufacturing facility.
vi. Any other pertinent EHS related information.

Health/medical surveillance is intended to monitor for biological effects of workplace exposures. It is intended to prevent and manage occupational illness and to promote health and productivity of employees. Employees who work with APIs (regardless of any PPE used) should be included in an appropriate periodic health/medical surveillance program. Health/medical surveillance records will be retained according to the local laws and regulations. As needed, the employees should follow the recommendations of the occupational health provider. After an accidental exposure, the employee should notify management and EHS professional, and if applicable, receive a post-exposure medical examination, which may include physical and/or clinical tests. Employees are strongly encouraged to report to supervisor(s) any adverse health reactions. Employee health issues or concerns will be treated discretely and confidentially. Women of child-bearing age are encouraged to notify their supervisor(s) or HR representative(s) of their pregnancy in order to assure safe work conditions are afforded.

Hazard Communication

The purpose of hazard communication is to inform employees of the hazards associated with the chemical and ensure that their work practices will not expose them to these hazards. The hazard communication means may include but not limited to the following: training, labeling, Material Safety Data Sheets (MSDSs), safety instructions in batch records/working methods, or special instructions. Global EHS develops MSDSs for APIs handled at company and makes them readily available to the sites. Site management should ensure employees are trained, at least annually, about the hazards, and the proper use of the required safety measures (including MSDSs). Training should also occur when a new material is introduced to the work environment.

Site management should ensure:

i. All APIs and excipients are received with supplier labeling and MSDSs.
ii. Warning signposts reflecting the assigned company OHC, or the appropriate safety instructions should be posted at the entrance to production suites/labs and on containers.
iii. Appropriate safety *instruction* relating to the OHC should be added to the batch card.
iv. Required employees attend medical surveillance and required EHS training. Unauthorized and untrained personnel are restricted from the area (Figure 12.1).

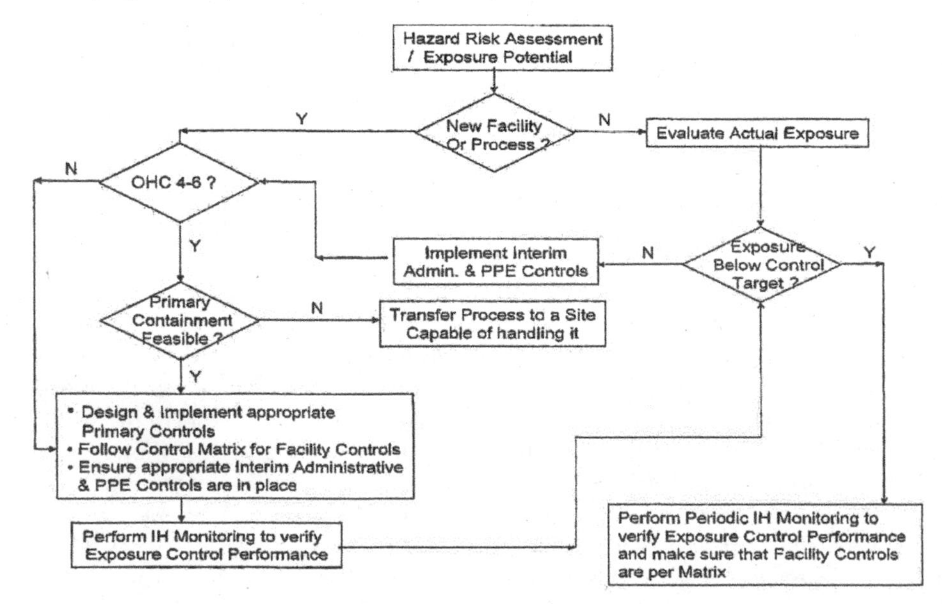

FIGURE 12.1 Hazard risk assessment flow diagram

INDUSTRIAL HYGIENE (IH) MONITORING GUIDELINES

Industrial hygiene monitoring is an essential part of the occupational exposure assessment of workers to APIs and validation of the effectiveness of exposure control measures. IH monitoring results are generally used to compare to the corresponding OEL or OHC according to regulatory or other requirements, but they may also be compared to other applicable values including a DEL, STEL, MUC APF, etc. While total inhalable dust monitoring of a process may satisfy local regulatory requirements, it may not provide an adequate certainty in comparison to the actual process and API and therefore monitoring of the specific API should be performed. Each site that handles hazardous chemicals, APIs, and physical and biological hazardous agents should implement an IH monitoring program.

This defines control targets that should be met and verified through IH monitoring. The IH monitoring program should follow the procedure outlined. The IH monitoring program may comprise the following elements:

- A walkthrough survey to assess the potential exposure hazards and a qualitative evaluation of exposure control effectiveness.
- A monitoring strategy to meet the appropriate monitoring targets including sampling location and duration, sample size, etc.
- The use of validated sampling and analysis methods.
- Interpretation, reporting and communicating results, conclusions, and recommendations, as required by local regulations and company policy.

The IH monitoring program must consider the relevancy of surrogate monitoring data when using it to compare to a specific API or multiple API's. The site manager will be responsible for executing the IH monitoring program, the preparations involved such as operational issues, availability of resources, and implementing the associated recommendations. The local EHS manager will be responsible for conducting walkthrough surveys, choosing the surrogate material monitoring planning, design, execution, and report authoring which includes data interpretation, conclusions, and recommendations. The local EHS manager should forward IH monitoring reports to corporate EHS for review and involve corporate EHS in IH monitoring activities as needed.

Definitions

Air Monitoring Methods – Company database of analytical methods for air sampling and analysis.

Assigned Protection Factor (APF) – The expected workplace level of respiratory protection that would be provided by a properly functioning respirator or a class of respirators to properly fitted and trained users. It is defined as the ratio of a contaminant concentration outside respiratory protecting equipment (RPE) to its concentration inside the RPE.

Design Exposure Limit (DEL) – The targeted exposure limit that is used for design purposes.

Excursion Limits – Where Short-Term Exposure Limit (STEL) is not applicable, a worker may be exposed up to three times the OEL for 30 minutes but under no circumstances should the exposure exceed five times the OEL at any time provided the OEL is not exceeded (ACGIH definition) Exposure Control Target for extended work schedules. The reduced airborne API exposure limit under which the exposure is acceptable for extended work shifts beyond 8 hours. The OEL or geometric mean of the OHC needs to be lowered accordingly for extended work shifts beyond 8 hours using the Brief and Scala Method shown below. This method *Daily Reduction Factor* = {8/h × (24-h/16) where h = hours worked per day *Adjusted Exposure Limit =8 hours. OEL × Daily Reduction Factor.* Therefore, the OEL or Geometric Mean of the OHC would be reduced 50% for a 12-hour work and 30% for a 10-hour work shift.

Highly Active/Potent Compound – Is an OHC 4, 5, or 6 compound Maximum Use Concentration (MUC) – The maximum concentration of a contaminant a person with respiratory protection can be exposed to safely and not have any adverse effects. This is based on a potential consistent concentration of the contaminant for an 8-hour shift (e.g., MUC = OEL × 1000 [protection factor of PAPR respirator]).

Short-Term Exposure Limit (STEL) – (American Conference of Governmental Hygienists (ACGIH) Definition) A 15-minute TWA exposure that should not be exceeded at any time during a workday, even if the 8-hour TWA is within the TLV-TWA. The TLV-STEL is the concentration to which it is believed that workers can be exposed continuously for a short period of time without suffering from (1) irritation, (2) chronic or irreversible tissue damage, (3) dose-rate-dependent toxic effects, or (4) narcosis of sufficient degree to increase the likelihood of accidental injury, impaired self-rescue, or materially reduced work efficiency. "Similar Exposure Group (SEG) –A group of persons who experience exposures similar enough to allow the prediction of exposure of a" members of the group by assessing the exposure of any member of the group. SEGs may be established through similarity in a process or task being performed together with the exposing agent, work conditions, and exposure location in a workplace (originally defined by the American Industrial Hygiene Association [AIHA]).

Surrogate Monitoring – Surrogate monitoring is monitoring a chemical substitute (usually inert or of low toxicity) for evaluating exposure control performance and when analytical methods are unavailable for the specific API that is being handled. Surrogate materials should have the same or similar characteristics (formulation process, physical characteristics, etc.). Lactose, naproxen, mannitol, and riboflavin are typical surrogates.

Upper Confidence Limit (UCI) – The upper point value of a confidence interval of a given statistical parametric entity (such as the mean value). A confidence interval generates a lower and upper limit for that entity. The interval estimate gives an indication of how much uncertainty there is in an estimate of the true value of the given entity. The narrower the interval, the more precise is the estimate.

IH monitoring is carried out to evaluate exposure potential, assess compliance with control target concentrations, and validate control measures applied. IH survey will be conducted in order to identify, recognize, and prioritize chemical, physical, and biological hazards and plan a comprehensive and appropriate monitoring program. A certified industrial hygienist shall conduct or direct the survey by the following steps:

i. Review unit operations and work processes.
ii. Review job and task description.
iii. Conduct qualitative/subjective observations of work methods and work practices.
iv. Identify and recognize APIs and their hazards, physical form, amount, and mass percent.
v. Define exposure profile/pathways and routes of body entry.
vi. Review frequency and duration of exposure.
vii. Review non-routine activities and situations leading to non-routine patterns of exposure.
viii. Review control measures (engineering, administrative, and PPE).
ix. Review relevant existing monitoring data.
x. Establish similar exposure groups (SEG).
xi. Prioritize the hazards when required.

Following the survey, a sampling strategy will be selected out of the following strategies to achieve monitoring objectives:

i. Baseline monitoring (obtaining a long-term exposure profile)
ii. Worst-case scenario (assessing top percentile of the exposure with a selected confidence level)

NOTE: Worst-case scenario may also refer to a work pattern where exposure determinants such as workload, physical effort, duration and frequency of exposure, expected airborne levels of the hazard, inferior protective measures, etc., contribute to the formation of top exposure.

Control evaluation/verification, surrogate monitoring, diagnostic monitoring (monitoring designed for in-depth investigation of a specific exposure scenario), where appropriate, sampling strategy will also address one or more of the following exposure patterns:

i. Shift or partial shift sampling (results compared to OEL or OHC).
ii. Short-term exposure sampling (results compared to STEL, MUC, or other recommended excursion limits).
iii. Exposure peaks for agents assigned with ceiling or IDLH values.

The monitoring program will comply with regulatory requirements. IH monitoring should reflect representative operations, typical and extraneous activities, workload, etc. in order to assess a representative exposure scenario. The timing of the monitoring should reflect a normal and routine schedule of work plans and work operations and activities. IH monitoring should be coordinated between site managers and EHS personnel so that all facilities, equipment, and other resources needed to assist the monitoring process are available at the time when the monitoring survey is conducted. Determine the type and number of samples. Select the appropriate validated IH method. In addition to published validated methods for regulatory compliance, company has developed a database of airborne monitoring methods to allow the selection of an appropriate method for monitoring a given API. Surrogate monitoring should be used in specific situations such as when a method is not available or for exposure control performance verification according to ISPE guidelines. Define sampling duration for each sample. Minimum sampling times are derived from analytical parameters in the specified method such as Limit of Detection (LOD) and Limit of Quantitation (LOQ) and sensitivity in order to eliminate erroneous results. When a mixture of APIs is used in a given facility, IH monitoring will ensure proper assessment of all components of the mixture. (Additive assumption should be used when two or more APIs have a similar pharmacological or toxicological effect on the same target organ or system; for example, if TWA of API 1 is 0.8 times the OEL and TWA of API 2 is 0.8 times the OEL, the employee's total exposure would be 1.6 times the OEL.) The measurements will be carried out or directed by a certified hygienist. Monitoring will follow local safety procedures.

Surrogate Monitoring

Select an appropriate surrogate (see addendum below). In many cases, the surrogate is used to ensure engineering control adequacy/performance such as prescribed by the International Society of Pharmaceutical Engineers (ISPE) guideline for assessing particulate performance for glove boxes, split valves, and other containment devices. Surrogate may be used for APIs with OHC or for other APIs with no adequate analytical method monitoring for the actual API or API's (worst case) may also be conducted as needed and no changes in PPE or controls are performed until adequate certainty is achieved with the monitoring results. Surrogates may be used when handling the API in an unverified containment device may be potentially hazardous and should be evaluated.

Data Interpretation and Handling

Analysis and interpretation of results should follow current and accepted IH procedures described and published in professional documents. It is preferable to use professional judgment in conjunction with descriptive statistics in order to optimize data interpretation and conclusions. Data interpretation and conclusions regarding the effectiveness of exposure controls and PPE are derived using the OEL or geometric mean of the OHC and in some cases protection factors such as APF and MUC. The monitoring results will be reported using a standard format designed by and approved EHS.

The report will include short background and description of the IH monitoring plan and objectives, presentation and interpretation (professional judgment and statistical analysis, etc.), and conclusions and recommendations (including recommendations for reevaluation). Recommendations should be tracked to completion. Report shall be retained according to the local laws.

Further Notes

Using point estimates and confidence limits for testing compliance of results with control limits:

1. An adequate level of certainty is required for data interpretation which is 95% or 99% of results below a point estimate with 95% or 99% confidence (UCL).
2. The 95% of a data set is the default choice for an adequate level of certainty; however, the 99% may be required in situations that require an added level of certainty, for example, reduction of PPE.
3. If the 95%- or 99%-point estimate is above the DEL, MUC, STEL, OHC, OEL, APF, etc., that number should be used for reporting/comparison purposes with the notation regarding the 95% or 99%.
4. If both the 95%-point estimate and the 95% UCL are below the DEL, MUC, STEL, OEL, OHC, APF, etc., the 95%-point estimate should be used for reporting/comparison purposes.
5. If the 95% is below the OEL, APF, etc., but the 95% UCL is above the DEL, APF, etc., then Bayesian statistics can be used with professional judgment in lieu of obtaining additional samples
6. Additional sampling may be performed in rare cases for clarification if the Bayesian statistical analysis still did not provide adequate certainty.

All values that are used for IH monitoring results such as DEL, MUC, STEL, DEL, OHC, and APF except for OHC are point estimates. The OHC is a range of exposure where the upper limit of acceptable exposure is the geometric mean of the range. In the absence of an OEL, the exposure profile shall be compared to the geometric mean of the OHC range except for OHC 6 which has no lower bound and therefore the upper bound shall be used.

Using assigned protection factors (APF) for evaluating PPE effectiveness is described as follows: when using exposure results to determine the adequacy of PPE, the airborne concentration can be divided by the APF. For example, if the OEI is 100 mg/mL and the protection factor is 10, then the user will be protected against concentrations up to 1,000 mg/mL which in that case is the MUC.

Protection factors together with the MUC may assist in selection of an appropriate PPE for a given exposure value. The minimum protection factor required to control a given exposure up to an acceptable one is obtained by dividing the actual exposure level by the OEI or the geometric mean of the relevant OHC range. The resulting value should be lower than the APF of the selected PPE. For example, given an actual respiratory exposure value of 1,000 mg/m^3 and an OEL of 5 mg/m^3, the selected PPE should have at least a protection factor of 200. In that case, a SAR or a PAPR would supply adequate protection.

Surrogate Selection – A material chosen as a surrogate for the measurement in order to simulate work with a given exposure material is required to represent properties and usage features which are like the exposure material in a given process or activity according to the following criteria:

- Particle size (surrogate may be milled).
- Uniformity of shape.
- Physical properties such as density.
- Chemical, physical, and environmental stability.
- Flow characteristics in the air and in process.

The surrogate material amount of use will be like this of the exposure material.

References

International Society for Pharmaceutical Engineering (ISPE) guideline for assessing particulate performance for glove boxes, split valves, and other containment devices.

American Industrial Hygiene Association (AIHA) strategy for assessing and managing occupational exposures.

International Society of Pharmaceutical Engineers (ISPE) Baseline Guide for Oral Solid Forms

CLEANING OF GLOVE BOXES AND BIOSAFETY CABINETS

This provides instructions on operating, cleaning, and general service of vertical flow Class II Type B1, B2, or B3, and Class III biosafety cabinets.

Definitions

BSC – Biosafety cabinet

Class 1 BSC – Open sash cabinet with downward airflow and HEPA filters.

Class III BSC – Commonly referred to as a glove box. Totally enclosed with gas- tight construction. Entire cabinet is under negative pressure, and operations performed through attached gloves.

Class II Type B1 BSC – A Class II BSC with 30% re-circulated cabinet air while exhausting the rest to the outside through HEPA filters.

Class II Type B2 BSC – Like Type B1, except that no air is re-circulated.

Class II Type B3 BSC – Re-circulate approximately 70% of cabinet air; other 30% is vented to the outside and ducts under negative pressure through HEPA filters.

PPE – Personal Protective Equipment.

HEPA – High-Efficiency Particulate Air.

Isolator – A type of glove box used for the weighing, dispensing, and compounding of API's.

Technicians – Any employee trained and qualified to perform maintenance, service, and certification on BSCs or a trained and qualified outside contractor.

Potent Compounds – A potent compound is an API or furnished product with an assigned OHC of 4, 5, or 6 and includes drug solutions, contaminated debris, and laboratory reagents containing potent drugs.

OHC – Occupational Health Category – A set of consecutive airborne concentration ranges (bands) used to classify the maximum permitted exposure levels for a range of APIs or finished drug products based on their health hazard, toxicological properties, and potency.

FACTS ABOUT BSC

A BSC is not intended for controlling vapors produced from organics compounds such as ether or chloroform. A BSC is not intended for controlling vapors produced from acids and bases such as sulfuric acid and sodium hydroxide. Engineering controls such as fume hoods are used to control vapors from these types of chemicals.

All Class II BSCs have vertically downward airflow and HEPA filtration. They are differentiated by the amount of air re-circulated within the cabinet, whether the air is vented to the room or to the outside, and whether contaminated ducts are under positive or negative pressure.

Class III BSCs are referred to as glove boxes. All air is HEPA filtered.

Each BSC is equipped with a continuous monitoring device that shows airflow pressure to allow confirmation of adequate airflow and cabinet performance.

Employees must properly use PPE and good work techniques for minimizing occupational exposure to hazardous drugs. Use of a BSC without proper use of PPE, training, and good work techniques may increase the risk of exposure.

PREVENTION OF EXPOSURE

To avoid contamination of the cabinet and of themselves, employee must use proper techniques when manipulating a hazardous drug. These prevention techniques include:

Minimization of personnel exposure:

I. Avoid inhalation of dusts, mists, or aerosols that are generated by agitation, dispersion, or pressurization of solids or liquid hazardous drugs, especially those techniques that are conducted in the breathing zone.
II. Prevent dermal absorption by using provided disposable impervious clothing. Remove, replace, and dispose of contaminated clothing immediately.
III. Wash hands after handling a hazardous drug to prevent contamination of food or cigarettes (e.g., ingestion). All PPE should be donned before work is started in a BSC.
IV. Assume aseptic techniques when working with hazardous drugs.

Minimizing Contamination of a BSC

I. Manipulations of a hazardous drug inside a BSC should not be performed close to the work surface of the cabinet.
II. Un-sterilized items, including liners and hands, should be kept downstream from the working area.

III. Entry and exit of the cabinet should be perpendicular to the front.
IV. All items needed should be placed within the SSG before work has begun. Extraneous items should be kept out of the work area.

PERSONAL PROTECTIVE EQUIPMENT REQUIREMENTS

Prior to handling a sealed container of a hazardous drug, the following personal protective equipment must be used by employee(s):

I. Two pairs of non-powdered latex/nitrile gloves.
II. ANSI-approved safety glasses with side shields.
III. Full-body Tyvek suit (bunny suit).
IV. Inner clothing consisting of disposable surgical scrubs (recommended).
V. Knee high Tyvek booties (shoe covers)

Note: Immediately dispose of contaminated clothing in appropriately labeled hazardous waste containers. Respiratory protection will be required when raising the sash for cleaning of a Class II BSC. All other manipulations of a hazardous drug inside a Class III BSC do not require the use of a respiratory device.

PRE-SAFETY CHECKLIST

Prior to using the BSC, check the continuous monitoring device to confirm adequate airflow or cabinet pressure.

Check to see if the BSC inspection and certification date are current. Located on the outside should be a sticker displaying the last date of calibration. All isolators and BSC II (glove boxes) must be calibrated twice a year. All BCS II and fume hoods must be calibrated annually.

Note: If the hood is out of calibration, tag the hood out of service and contact Metrology for calibration of cabinet. Examine the inside of the chamber, paying especially close attention to any deposits on the countertop, sides of the cabinet, and along the front edge of the BSC (Class II only).

Note: Deposits found must be cleaned following the cleaning and decontamination procedure.

CLEANING AND DECONTAMINATION

All BSC must go through a thorough cleaning at least annually or if a large spill occurs which contaminates more than 50% of the BSC's working surface. For most purposes, cleaning can consist of using a surfactant (detergent) to clean the inside of the chamber, followed by a water rinse.

Decontamination should consist of surface cleaning with water and detergent followed by thorough water rinsing. The use of detergent is recommended because there is no single accepted method of chemical deactivation for all hazardous drugs.

Quaternary ammonium cleaners should be avoided due to the possibility of vapor.

Avoid using ethyl alcohol or 70% IPA as a cleaner since these hoods are not explosion proof.

Spray cleaners should also be avoided due to the risk of spraying HEPA filters.

Ordinary decontamination procedures, which include fumigation with a germicidal agent, are inappropriate in a BSC used for hazardous drugs because such procedures do not remove or deactivate the drugs. Fumigation is primarily used for inhibiting microbial growth.

All BSCs should have in place a plastic-backed lining on the working surface which covers the interior surface of the chamber. The lining should be replaced daily.

The sash of a Class II BSC should remain down as far as possible during cleaning; however, a NIOSH-approved respirator equipped with a HEPA pre-filter and organic vapor cartridges should be used by the employee if the sash will be lifted during the cleaning process.

Note: Employees cleaning BSCs must be trained on SOP.

The exhaust fan blower should be left on when cleaning is being conducted.

Note: Avoid gross application of the cleaning solution to avoid wetting the HEPA filters.

Cleaning should proceed from the back of the BSC, away from the user. The drain spillage trough area should be cleaned at least twice since it can be heavily contaminated.

All materials from the decontamination process should be handled as hazardous and disposed of accordingly.

MAINTENANCE, SERVICE, AND CERTIFICATION OF BSCS

BSCs are to be serviced and certified any time the cabinet is moved or repaired.

If a cabinet has been moved or been repaired, notify the Metrology Department. Place a sign on the cabinet stating "Out of Service" along with a date, corresponding department name, and contact number. The Metrology Department will arrange for a contractor to recertify the chamber.

Contracted technicians servicing these cabinets or changing the HEPA filters must be informed by Metrology of the hazardous drug risks.

Contracted technicians servicing these cabinets should use the appropriate respiratory protection.

HEPA filters should be replaced when they restrict airflow, or if they are contaminated by an accidental spill. They should be bagged in plastic chemotherapy waste bag and disposed of as chemo drug debris.

REFERENCES

OSHA Instruction TED 1.15, September 1995. Office of Science and Technology Assessment, Section V: Chapter 3, Controlling Occupational Exposure to Hazardous Drugs.

FACILITY/INFRASTRUCTURE CONSIDERATIONS: LABORATORY OPERATIONS

This matrix intends to provide a general guidance for the required containment strategy. Choosing the appropriate facility consideration should be based on a professional case by case judgment of the exposure potential, considering the dustiness, quantities, and the task duration. When the given judgment goes against the matrix recommendations, the rationale of the chosen control strategy should be documented and approved by EHS in advance. Exposure Control Target is the OEL or the geometric mean of the relevant OHC range. Controlled containment is an engineered containment solution that always satisfies the Exposure Control Target, with a very low probability to fail. General considerations for this case apply only if all processes within the room (facility) are in controlled containment systems.

FACILITY/INFRASTRUCTURE CONSIDERATIONS: PRODUCTION AND PILOT PLANT OPERATIONS

This matrix intends to provide a general guidance for the required containment strategy.

Choosing the appropriate facility consideration should be based on a professional case by case judgment of the exposure potential, considering the dustiness, quantities, and the task duration.

When the given judgment goes against the matrix recommendations, the rationale of the chosen control strategy should be documented and approved by EHS in advance. Exposure Control Target is the OEL (if it has been established) or the geometric mean of the relevant OHC range. Controlled containment is an engineered containment solution that always satisfies the Exposure Control Target, with a very low probability to fail. General considerations for this case apply *only* if all processes within the room (facility) are in controlled containment systems.

CONTAINMENT CONTROL GUIDELINES/ENGINEERING MATRIX: LABORATORY OPERATIONS

Comment

This engineering matrix intends to provide a general guidance for the required containment strategy. Choosing the appropriate engineering containment solution should be based on a professional case by case judgment of the exposure potential, taking into account the dustiness, quantities, and the task duration. When given judgment goes against the matrix recommendations, the rationale of the chosen control strategy should be documented and approved by EHS in advance. "Engineering Toolbox" attachment may assist in selecting the appropriate solution out of the available solutions in the market.

Terms and abbreviations used in this table:

Isolator – Glove box or flexible glove bag
LEV – Local exhaust ventilation
BSC – Bio-safety cabinet

CONTAINMENT CONTROL GUIDELINES/ENGINEERING MATRIX: PRODUCTION AND PILOT PLANT OPERATIONS

COMMENT

This engineering matrix intends to provide a general guidance for the required containment strategy. Choosing the appropriate engineering containment solution should be based on a professional case by case judgment of the exposure potential, taking into account the dustiness, quantities, and the task duration. When given judgment goes against the matrix recommendations, the rationale of the chosen control strategy should be documented and approved by EHS in advance. "Engineering Toolbox" attachment may assist in selecting the appropriate solution out of the available solutions in the market.

Terms and abbreviations used in this table:

LEV – Local exhaust ventilation
SBV – Split butterfly valve
FBD – Fluidized-bed dryer
Isolator – Glove box or flexible glove bag
Centrifuge Types: TD – Top discharge
BD – Bottom discharge
Hor. – Horizontal peeler
Inv. Basket – Inverted Basket

13 Data Integrity Compliance

Procedures for compliance with 21 CFR Part 11 – must adhere to data integrity compliance, e.g., all lab instruments, process control systems (PLC or HMI), computerized systems, and excel spreadsheets used should all comply with data collection, storage, and usage requirements such as:

1. PLC (programmable logic controllers) or other intelligent processors
2. HMI (human machine interface)
3. Analytical instrument software
4. Computer systems

Computer System – A combination of computer software (including quality excel spreadsheets), hardware, and peripherals designed to perform a specific function. Examples of computer system include process control (equipment, environment), business (accounting, enterprise resource management), and database (control charting, LIMS).

Critical Data – This (including parameters) is defined to be any data created, modified, or maintained by a system/instrument/PLC/HMI which will support quality decisions related to product safety, efficacy, and the quality of the product such as:

- Non-conformance investigation
- Modifications/maintenance activities for equipment
- Product release requirements
- PLC or HMI parameters that could potentially impact product acceptability.

Data Integrity – The extent to which all data are complete, consistent, and accurate throughout the data lifecycle. Complete, consistent, and accurate data should be attributable, legible and permanent, contemporaneously recorded, original or a true copy, and accurate (ALCOA).

Dynamic Records – A dynamic record is a record format that allows interaction between the user and the record content. Examples of a dynamic record include chromatographic data file, which allows processing to change the baseline, or a spectral data file, which allows changes in the displayed output.

GXP – General term used to represent multiple regulations defined by Regulatory Agencies (e.g., Good Manufacturing Practices (GMP), Good Clinical Practices (GCP), Good Laboratory Practices (GLP), etc.) that are applicable.

Information Management – The group responsible for management of a computer system.

DOI: 10.1201/9781003224716-13

Metadata – This is the contextual information required to understand the original or processed data. Metadata is often described as data about the data. It is an integral part of the original (raw) data or process data (without metadata, data has no meaning). Metadata for a particular piece of data could include the date/time stamp for when the data was acquired, a user ID of the person who generated the data, the instrument ID used to acquire the data, etc. An audit trail is a form of metadata.

Predicate Rule – Federal regulation that defines the requirements for record creation, retention, and/or signature/individual identification.

Static Records – A static record is a fixed-data document such as a paper record or an electronic image. Examples of a static record include a paper printout from a balance or a static image created during the data acquisition (Table 13.1).

21 CFR Part 11: Electronic Records – 21 CFR Part 11 defines the requirements for use of electronic records and electronic signatures which are used in lieu of paper records (21CFR Section 11.2) in which the records are required per GXP procedures. Any laboratory instrument, PLC, or computerized system which retains electronic records that are used to replace existing paper records are covered by the requirements of 21 CFR Part 11 for electronic records. Systems that maintain electronic records and the intended use is paper are not considered to be covered by 21 CFR Part 11, although the system must comply with all data integrity requirements.

21 CFR Part 11:– Electronic Signatures – 21 CFR Part 11 defines the requirements for use of electronic signatures associated with electronic records if the signature is used to replace a signature recorded on paper. Any laboratory instrument, PLC, or computerized system that is used to replace paper-based signatures are considered electronic record systems and must comply with all requirements of 21 CFR Part 11 for electronic signatures. Electronic signature

TABLE 13.1
Data Integrity: Roles and Responsibilities

Local Role	Responsibilities
Software quality engineer Quality lab associate	Responsible for overall management of the compliance program for areas assigned, maintenance of inventory list, leading coverage and gap analysis efforts, and remediation activities (if applicable).
Area manager Quality engineer manager/quality System representative	Responsible for approvals of coverage assessments for 21 CFR Part 11 as well as data integrity assessments upon completion.
Computer system owner/business Process owner	Responsible for assisting with coverage assessments for 21 CFR Part 11 and data integrity.
Information management	Responsible for assisting in inventory and analysis efforts and administration of remediation process as required.

training must be documented and verified prior to adding new users to an electronic signature system.

Data Integrity – Any laboratory instrument, PLC, or computerized system which retains GXP relevant data (regardless of Part 11 Coverage) must comply with all data integrity requirements to ensure data is complete, consistent, and accurate throughout the data lifecycle.

PROCESS/PROCEDURE: INVENTORY OF SYSTEMS

An inventory of all laboratory instrument, PLC, or computerized system will be maintained for all systems (laboratory, business, and process control) in use. The following information is required for each system:

a. System name
b. Functional description
c. Platform (MES, lab instrument, PLC, HMI)
d. Location
e. Business process owner
f. 21 CFR Part 11 coverage
g. GAMP category (category 1, 3, 4, or 5)
h. Records maintained required by GXP (Yes or No)
i. Types of GXP relevant data maintained by the system
j. Validation specification reference
k. Risk classification
l. System classification (category)
m. System risk classification (high, medium, low)
n. 21 CFR data integrity compliant
o. Gaps identified (if any)

PROCESS/PROCEDURE: DETERMINATION OF 21 CFR PART 11/DATA INTEGRITY APPLICABILITY

Process/Procedure: 21 CFR Coverage Assessment

All laboratory instruments, PLC, or computerized systems must be evaluated and classified for Quality System impact, 21 CFR Part 11, and Data Integrity applicability. The results will be recorded in the computer system inventory.

Process/Procedure: 21 CFR Part 11/Data Integrity Gap Analysis

Any laboratory instrument, PLC, or computerized system that is covered by 21 CFR Part 11 must complete a gap analysis to determine if the system is fully compliant with all requirements. Any gap identified must be addressed procedurally or with a system change to bring the system into full compliance prior to implementation (Table 13.2).

TABLE 13.2
Change History

Change	Description of Change
	Add excel spreadsheet clarifications and new sections for coverage assessment and gap analysis.
	Combine forms X and Y into Z form. Also update inventory elements as required.
	All sections updated for data integrity requirements. For example, formatting change only. Obsolete form X. Created new form for data integrity evaluation Y. Re-process: Form Z, added definition of critical data – alphabetized definitions. Form Z – added additional questions for assessment related to autosave and security.

14 Guidelines for Statistical Procedure

This is to establish guidelines for the use of statistical techniques in protocols and other official studies (Schmitt [61]).

1. Alpha (a) The probability of rejecting a true hypothesis (typically referred to as producer's risk)
2. Beta (B) The probability of accepting a false hypothesis (typically referred to as customer's risk)
3. DPM Defects per million
4. CPM Complaints per million
5. USL Upper specification limit
6. LSL Lower specification limit
7. MSA Measurement system analysis

Basic guidelines (Torbeck [87]) will be set forth for each procedure; however, the instruction for a procedure may not contain enough information to execute. If this is the case, a statistics text such as the *Quality Control Handbook* should be consulted. Minitab statistical software may be used for analysis. If Minitab is used and data is not stored on the server, the data will be included with the study results (Torbeck [83]).

Hypothesis testing involves procedures that compare two groups to identify if there is a statistical difference between them (Torbeck [84]). This procedure will cover three different hypothesis tests:

1. Z test – comparison of means, standard deviation known
2. T test – comparison of means, standard deviation unknown
3. F test – compares the variance of two samples

The basic assumption or null hypothesis used in hypothesis testing is that the samples are from the same population (Torbeck [74]). The purpose of the test is to determine if there is enough evidence to state that a statistical difference exists. The procedure cannot prove two samples are the same only that there is not a statistical difference. Hypothesis testing parameters are selected based on two types of risks:

1. Alpha (a) The probability of rejecting a true hypothesis (typically referred to as producer's risk)
2. Beta (B) The probability of accepting a false hypothesis (typically referred to as customer's risk)

DOI: 10.1201/9781003224716-14

The alpha and beta parameters should be chosen based on the criticality of the analysis and the effect of each instance occurring. Reducing the parameter enhances the sensitivity of the test to that occurrence. As a general guideline, alpha is normally set at 0.05 and beta at 0.10. These parameters are used to determine the acceptance region for the test. Tables for this value are found in most statistics texts or *Quality Control Handbook.*

A one- or two-tailed evaluation must be specified to choose the acceptance region. A one-tailed test is used when the average must fall on one side of the mean. If the test results must be either higher or lower than the current process, a one-tailed test is used. The two-tailed test is used if the test mean may fall on either side of the control mean. The number of tails is important when choosing the acceptance region for the test. A two-tail test will require the alpha level to be divided by two to capture equal areas on either side of the mean.

Defects per million (DPM) or complaints per million (CPM); complaint incidents per million (CIPM)

DPM is calculated by the following formula:

Defects/units produced × 1,000,000

CIPM is calculated by the following formula:

Complaint incidents/units produced × 1,000,000

It should be noted that DPM cannot be added. To recalculate, simply add all defects or complaint incidents together and divide by the sum of the total units produced.

PROCESS CAPABILITY ANALYSIS

A process capability analysis is used to predict process performance. This study is powerful as it allows the use of variable data to better predict process performance. The analysis allows predictions to be made on the process or product. The example Figure below is a guide as a normal distribution model to understand process performance (Figure 14.1).

By using the above reference, we can determine that the bell-shape distribution would cover 99.73% of the data. The location and the spread can easily determine the capability of the process. The examples below give the histogram when used with upper and lower specification limits (Figure 14.2).

Process capability refers to the natural variation of a process which occurs due to common causes. This is calculated by comparing a measure of the variation to the specification limits for the characteristic of interest. In this way, the ability of a process to meet the product requirements can be quantified. There are two types of studies: process potential (Pp) and process performance (Ppk). Pp and Ppk may be used to qualify a process with variable data. The Pp measure is a ratio of the process variation to the specification limits. Pp is normally performed under optimal conditions and is short in duration. The data is typically not maintained in a time order. The Ppk measure accounts for process centering and stability. Pp studies are

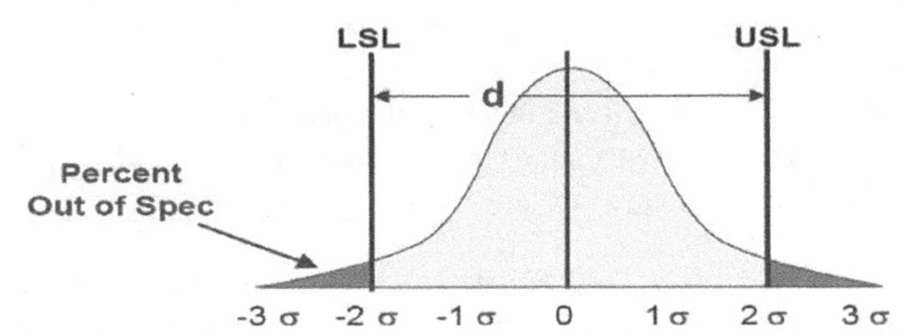

Distance From Average	Percentage Within	Percentage Outside
±0.5σ	38.3%	61.7%
±1σ	68.3%	31.7%
±1.5σ	86.6%	13.4%
±2σ	95.4%	4.6%
±2.5σ	98.8%	1.2%
±3σ	99.7%	0.27% or 2700/million
±3.5σ	99.95%	465/million
±4σ	99.994%	63/million
±4.5σ	99.9993%	6.8/million
±5σ	99.99994%	0.6/million

FIGURE 14.1 Distribution model.

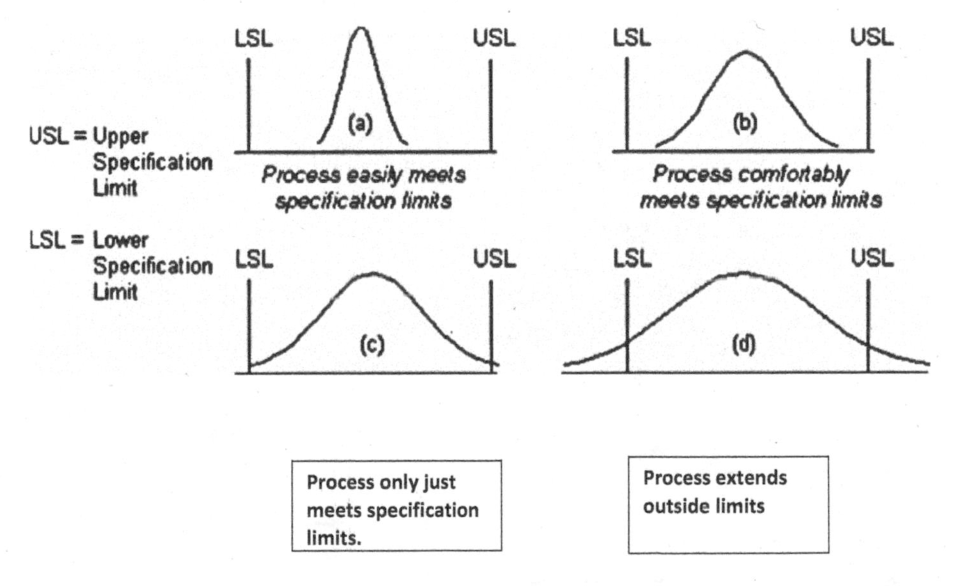

FIGURE 14.2 Specification limits.

appropriate for short-term (consecutive) testing. Note Ppk can be used to review the processes centering ability.

1. Study may be performed during actual production or under conditions that simulate actual production.
2. For a snapshot of the process capability, a minimum of thirty samples can be produced continuously. Any sample size greater than thirty is acceptable as it improves the data reliability. No process adjustments will be made, and the same operator will be used.

3. Measure the characteristic of interest using a calibrated instrument(s) and record this number of the data sheet.
4. Calculate Pp and Ppk. Desired Pp and Ppk measurement is 1.33 with not out of spec values. Note other variable plans may be used, such as Mil Std 414 or other variable sampling methodologies.

Note: Central Limit Theorem – “For most populations, the sampling distribution of the mean can be approximated closely by a normal distribution, provided the sample size is sufficiently large.” “Approximate normality is clearly evident at $N=30$.”

Minitab will present the overall capability as Pp and Ppk. The overall capacity (Pp and Ppk) is to be used for all studies with a sample size of one in which the data is not tested in sequence. This overall evaluation calculates the standard deviation based on all samples.

LONG-TERM STUDIES

Cpk studies are long-term studies, which samples are taken at different time periods. These studies should be performed as follows (Figure 14.3):

1. Study may be performed during actual production or under conditions that simulate actual production.
2. Samples should be pulled at regular intervals over a specified period to yield a true process picture. Example: Pull samples every hour for three shifts of production.
3. Measure the characteristic of interest using a calibrated instrument(s) and record this number of the data sheet.
4. For long-term studies, place the 50 samples into ten subgroups of five units each. A standard subgroup size is five samples.

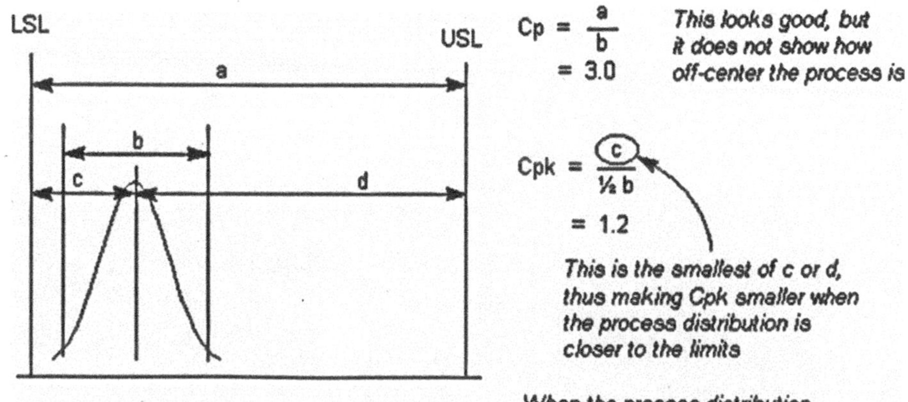

FIGURE 14.3 Process capability.

5. Calculate the Cp and Cpk of the process.
6. Review the control chart for points out of control or out of specification.
7. Desired Cp and Cpk measurement is 1.33 with not out of spec values. Note: Other variable plans may be used such as Mil Std 414 or other variable sampling methodologies.

Minitab may be used and is the preferred method for performing capability and gauge repeatability and reproducibility studies. Minitab will also present the within capability study results as Cp and Cpk. These calculations are based on longer-term studies in which the data is collected in time sequence. These studies typically use averages subgroup's range or subgroup's standard deviation to calculate the overall standard deviation. These within calculation from minitab (Cp or Cpk) will only be used with time sequence-based data.

ACCEPTANCE SAMPLING

Attribute and Variable Sampling Plans

Sample plans are to be selected based on AQL and LTPD, in order to replace a current plan, follow the steps outlined below. The AQL/LTPD should be used for finished product. Component level or in-process sampling plans should be tighter than those used for finished product.

1. Calculate the AQL and LTPD of the current plan.
2. Find other methodologies that have similar AQL's and LTPD's.
3. Choose the lower cost or easiest to maintain.
   LTPD – Lot Tolerance Percent Defective
   AQL – Acceptable Quality Limit

Sampling plans may be calculated using Wayne Taylor's sample size software or other standards.

Variable Sampling Plans: ANSI Z1.9

Variable sampling plans may be substituted for attribute plan provided they represent the AQL's and LTPD's needed. Variable plans use the sample size and a K constant to establish confidence levels. The average and standard deviation from the analyzed data is compared to the K values to determine acceptance. The K value represents the number of standard deviations the mean is from the limit.

The formula for calculation is listed as follows:

$$(\text{USL} - \text{Average})/\text{Standard Deviation} = X$$
$$(\text{Average} - \text{LSL})/\text{Standard Deviation} = y$$

The minimal value of X and Y must be greater than K (standard) for the data to be acceptable.

Normality

The normality assumption must be assessed when using variable sampling plans with standard deviation unknown for non-standard processes. The method to test for normality can be selected from the list below and should be conducted using sample sizes of minimum 30 units and a confidence level of 90%. These normality tests are available in Minitab.

1. Probability plotting of individual values and achieving an adequate linear fit of the data.
2. Performing a statistical test of normality, such as:
   a. A Shapiro-Wilk test.
   b. A Chi-Square test.
   c. The Anderson-Darling test for normality.
   d. Any test having an equivalent statistical power to the previous tests.

The test rejects the hypothesis of normality when the p-value is less than or equal to 0.1.

Failing the normality test allows you to state with 90% confidence the data does not fit the normal distribution. Passing the normality test only allows you to state no significant departure from normality was found.

The following graphs display normal distributed data and an example of the application of the Anderson-Darling (AD) test for normality. The test confirms that the data meets the normality assumption at a confidence level of 90%.

No data perfectly matches a normal distribution. The normal distribution is typically more conservative than other models and therefore is enough for most analysis (see reference below).

"Three sigma limits do, indeed filter out virtually all of the routine variation and they do this regardless of the shape of the model used. This means that we do not have to have normally distributed data for the limits to work. Three sigma limits are completely general and will work with all types of routine variation. By filtering out virtually all of the routine variation they will yield very few false alarms."

Reference "Normality and the Process Behavior Chart" Donald J Wheeler SPC Press

Attention will be paid to the tails of the data and potential outliers (>4 standard deviations). There is a reasonable chance that points may be beyond the 3 standard deviation limits of a normal distribution, especially as the sample size increases. For example, 50 samples from a normal distribution have a 13.5% chance $((100\% - 99.73\%) \times 50)$ that a point will lie outside 3 standard deviations.

Outliers are not to be removed unless the cause is identified, and the testing is repeated.

Under unique circumstances, data may yield a distribution that will not fit the normal distribution assumption. Extreme deviations from the normal distribution may require normalization or other statically sound methods.

The following graph is an example of data that could require transformation or other statistically analysis to be able to predict confidence levels (Figure 14.4).

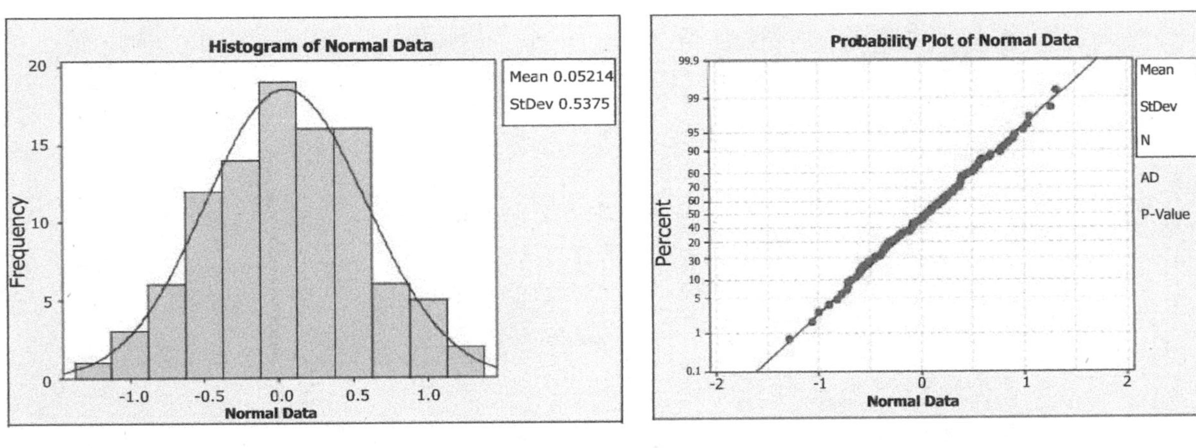

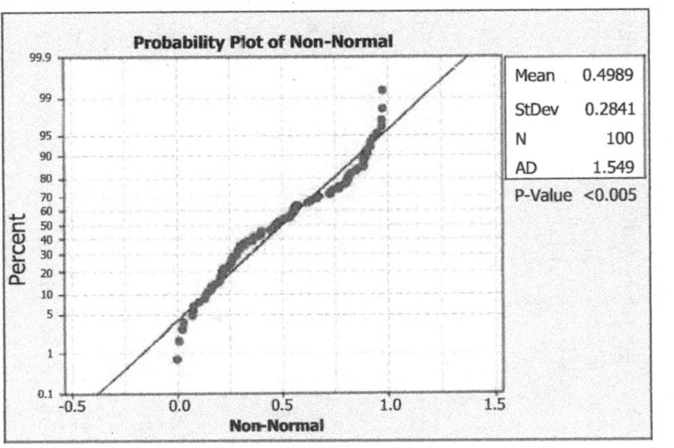

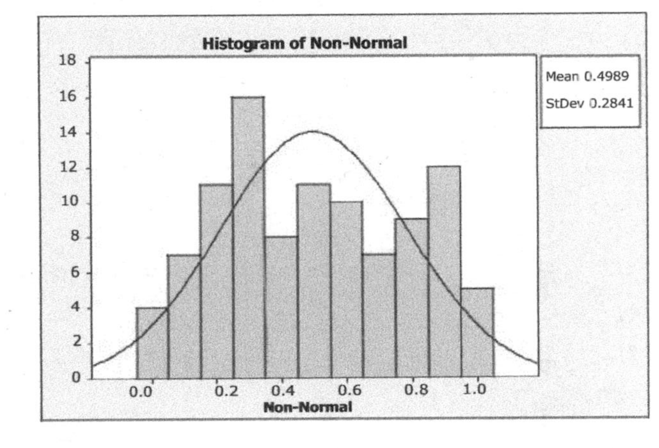

FIGURE 14.4 Normality distribution.

TRANSFORMATION OF NON-NORMAL DATA (NORMALIZATION)

If the data to be analyzed are not normally distributed, transform the data by taking the natural logarithm of each value. Then perform a second normality test on the transformed values to verify that the data behave normal. If the transformed data are not yet normally distributed, consult a person trained on statistical data analysis (i.e., a Black Belt, or a statistician) to review the data and provide rationale for the analysis performed.

PROTOCOL SAMPLING

Acceptance sampling will be applied in two different ways during a protocol. During the IQ or OQ phases, a zero accept plan may be used to qualify a process or a production lot while a 90% LTPD plan may be used in PQ to evaluate finished goods batches.

A zero accept plan is chosen based on the AQL for the quality characteristic(s) affected. These plans ensure a 95% confidence that the process defect level is at or below the specified AQL. Samples may be produced continuously or pulled randomly (Table 14.1).

This table was developed using the formula 3 (constant)/AQL expressed as fraction defective. This formula may yield slightly higher sample sizes. A more precise method is to use Sampling Software version 1.0 or higher from Taylor Enterprises.

LTPD plans to be used in the PQ phase are modified versions of ANSI Z1.4 double sampling plans. The accept reject levels have been modified to achieve an LTPD equal to the desired AQL. These plans will be selected independent of lot size based on the AQL of the defect(s). The stated AQL and LTPD of the plan is based on a process and not individual lots (Table 14.2).

FAILURE MODE AND EFFECT ANALYSIS (FMEA)

For the assessment and management of risk, the use of analysis techniques is recommended. They will serve as prevention tools to manage and reduce risks associated

TABLE 14.1
Zero Accept Plan (Typical OQ Plan)

AQL (%)	Accept	Minimum Sample Size
0.065	0	4,615
0.10	0	3,000
0.15	0	2,000
0.25	0	1,200
0.40	0	750
0.60	0	500
0.65	0	462
1.00	0	300
1.50	0	200
2.50	0	120
4.0	0	75

TABLE 14.2
90% LTPD Plan (Typical PQ Plan)

Defect(S) AQL	Sample 1	Accept/ Reject	Sample 2	Accept/ Reject	AQL p(a) = 0.95	LTPD p(a) = 0.10
< 0.065	3,850	0/2	3,850	1/2	0.054	0.0647
0.65	385	0/2	385	1/2	0.054	0.653
1.0	385	1/3	385	2/3	0.14	1.02
2.5	150	1/3	150	2/3	0.35	2.59

with the failure of machines, processes, and systems. Techniques like Fault Tree Analysis (FTA), Hazard and Operability Analysis (HAZOP), Failure Mode and Effects Analysis (FMEA), or other sound method may be used.

The goal of an FMEA is to define and eliminate problems prior to or after implementation. The tool may be applied to machines, product, process, computer system, etc. This risk analysis will be applied to Change Control processes. The Quality Engineer will determine if an update of the FMEA is required based on input from the change owner.

The standardized Risk Priority Number (RPN) approach will allow different processes to be compared in the plant. The FMEA may also define key risk areas that require corrective action or increased validation. The tool is a commonsense approach to risk identification based on SME empirical knowledge and experience. The occurrence of the defect is related to the criticality and ability to remove the defect. As each step in the process is reviewed, a risk value is assigned based on the occurrence times criticality times inspection level.

An attachment to the FMEA will be created and will include the following items:

1. The process steps will be recorded if applicable.
2. The potential failure mode will be listed for each process step. Multiple failure modes for each process step may exist.
3. The potential cause for the failure mode will be identified. Multiple causes for the failure mode may exist.
4. The potential effect of the failure will be listed. If available, this should align with the AQL defect classifications.
5. **Occurrence** – Chance of the potential cause occurring and resulting in the failure mode. A rating of 1–5 will be assigned. Use scrap rates and other applicable information if known.
6. **Severity** – Severity of failure. A rating of 1–5 will be assigned. The product specifications should be used if applicable. Special attention is to be given to scores or 4 or 5 (quality items).
7. **Detection** – Detection rate of failure. A rating of 1–5 will be assigned. A review of product checks, machine controls, and downstream inspection should be included in this review.
8. **Current Controls** – A listing of current control may be listed or referenced from the process flow chart if provided.

TABLE 14.3
RPN Values

RPN (O×S×D)	Comments and Rationale Required?	Approvals Required
≥ 45	Yes	Routine approvals, quality manager, quality director
28–44	Yes	Routine approvals, quality manager
13–27	Yes	Routine approvals
1–12	No[a]	Routine approvals

[a] For failure modes with a severity rating of 4 or 5, comments or rationale are required. When developing FMEAs for new products, processes, or equipment, all failure modes with a severity of 4 or 5 must be reviewed to ensure that some control or detection is in place.

9. **RPN** – Calculate the RPN. Multiply Occurrence x Severity x Detection to determine RPN. High RPN ratings need to be addressed to minimize risk and that documentation reflects the corrective action(s). Reference the table below for comments, rationale, and approval requirements related to RPN values (Table 14.3).

Comments: List comments, rationale, and/or corrective actions related to the step, if needed. Reference table above to determine when required (Table 14.4).

Applicable approvals must be obtained from appropriate areas as related to the FMEA. Approvals indicate agreement with the analysis. For FMEAs with RPN values 28–44, approval from the area affected quality manager is required. For FMEAs with RPN values ≥ 45, quality manager and quality director approval is required. After the document has been assigned an FM number, the FMEA has been created and approved, and QE will update the FMEA master file.

Environmental FMEAs are used for risk assessment of controlled environmental rooms and processes. Environmental FMEAs are performed and documented and are controlled in addition to the SOP change control process.

Calculating or Recalculating Control Limits

Control limits are special upper and lower limits used to monitor the performance of a process. These limits are calculated from actual process data and are generally

TABLE 14.4
An Example of an FMEA Format

1) Process Step	2) Potential Failure Mode	3) Potential Cause	4) Potential Effect of Failures	5)	6)	7)	8) Current Controls	9) RPN O×S×D	10) Comments

positioned three standard deviations from the process mean. Theoretically, this range of + or − three standard deviations will contain 99.7% of the process data; therefore, any data point outside these limits is indicative of a significant process shift. Limits must be reviewed periodically to detect changes in process performance which may require narrowing or widening the limits.

Procedure

1. Limits are to be calculated based upon the controlling plant procedure.
2. Limits should be established based on historical data with the limits calculated for ±3 standard deviations. Limits may be derived with a spread other than ±3 standard deviations but these exceptions must be noted, and the reason documented in the applicable SOP.
3. Limits may not be established outside of spec limits. If the control limits are greater than or equal to the spec limits, control limits cannot be used for that process.
4. Limits should be recalculated using a minimum of 30 data points pulled randomly from historical data. If available data for the period reviewed does not contain more than 30 data points, then all data should be used.

15 Calibration

Routine calibration of equipment and documentation of the calibration results is outlined by reflecting on methods of calibrating and documentation practices of miscellaneous devices, which measure products or processes. All inspection, measuring, and test equipment that can affect product quality or processes are to be calibrated against Standard Measurement and Test Equipment (M&TEs), which are traceable to the National Institute of Standards and Technology (NIST). If NIST or national traceable standards are not available for the parameter being measured, an independent reproducible standard shall be used. The following is to establish approach to Control of Measuring and Test Equipment-Calibration Program.

Calibration – Set of operations that establish, under specified conditions, the relationship between values of quantities indicated by a measuring asset or measuring system, or values represented by a material measure or a reference material, and the corresponding values realized by Standard M&TEs.

CMMS – Computerized Maintenance Management System is a centralized computer system controlled by client's global IT department, with all system servers residing in client's outsourced data center. CMMS is comprised of a web-based computer enterprise planning system application developed by IBM that can be accessed by any computer in the client network using Internet. All asset records are maintained electronically using computer enterprise planning system. Computer enterprise planning system is also used for all maintenance records.

Measurement Standard/M&TE (Measurement and Test Equipment) – Material measuring asset, reference material, or measuring system intended to define, realize, conserve, or reproduce a unit or one or more values of a quantity to serve as a reference.

Test Accuracy Ratio (TAR) – Ratio of the accuracy of the unit under test (UUT) and the reference standard used to calibrate the UUT. A test accuracy ratio of 4:1 is required for measurement standards where TAR can be applied.

Test Uncertainty Ratio (TUR) – Ratio of the accuracy tolerance of the UUT to the uncertainty of the measurement standard (M&TE) used.

Tolerance – Extreme values of an error permitted for a given asset.

As new assets are received, they will be reviewed by plant engineering and quality management departments to determine any calibration requirements. For replacement of existing equipment, area QM supervisor or appropriate manufacturing

DOI: 10.1201/9781003224716-15

supervisor should always notify the calibration supervisor or calibration technician as these assets are obtained. Calibration requirements are reviewed as follows:

1. Analyze the use of all assets to determine whether the asset should be calibrated.
2. It is the owner's responsibility to assure that the calibrated equipment is suitable for the application in which it will be used. Suitability includes assuring that the range; tolerance, accuracy (uncertainty), and capacity of the asset/equipment are enough for the application.
3. Evaluate manufacturer's recommended calibration process and provide calibration procedures in conjunction with the calibration department.
4. Whenever an asset Out of Tolerance (OOT) Report is assigned, the owner will notify all users of the asset since its last calibration and will complete the OOT Report.
   a. It will be the responsibility of the owner and/or area supervisor to notify the calibration lab of new assets and changes in usage to any assets. When an operation limit or calibration frequency is changed, documentation to support that change will be recorded on the calibration work order (i.e., protocol, critical work numbers, or rationale for the change) and approved by quality management.
   b. Obvious changes to the operation limit such a temperature or mass unit designations (i.e., kg to lb.) will not need supervisor approval.
   c. All assets being made inactive or decommissioned must be turned in to the calibration lab for a final verification and status update in the calibration system.
   d. It will be the responsibility of the owner and equipment users to assure that the equipment is used only within the calibrated range, that assets are not used beyond the calibration expiration date, and that all equipment is properly identified.
   e. Provide calibration procedures.

Calibration Lab Responsibilities

1. Perform calibrations
2. Maintain NIST Traceable Standard M&TEs. The calibration lab must always remain locked to limit access to M&TEs for security and to prevent damage of standards.
3. Review and file calibration paperwork generated by the calibration lab.
4. Issue OOT Reports, as required.
5. Maintain the records in computer enterprise planning system.
6. Generate a daily overdue calibration report.
7. Generate a daily active asset/active PM report.
8. Generate weekly computer enterprise planning system reports for supervisor review.
9. Generate monthly calibration reports
10. Issue asset status notification forms as required.

Assets calibrated will show an accuracy ratio of 4:1 or better. The test ratio is determined using the manufacturer's stated accuracy of the equipment being used against the process tolerance of the UUT. A list of temperature calibrations performed by the calibrations department and all the equipment available to perform the calibration are maintained and updated as needed. It shows that all equipment maintained in the calibrations department can be combined in any combination without dropping below the required 4:1 ratio. Procedures will be used for determining manufacturer accuracies and the uncertainty or combined uncertainty of any component(s) used to perform a calibration. Procedures are required to be updated whenever changes occur to Standard M&TEs or when adding or deleting a Standard M&TE from service.

Hard copies of instrumentation manuals, showing where all accuracies for all Standard M&TE equipment were derived, are filed in the Calibrations Lab in the Manuals Section of the Standards File Section.

Note: The Test of Uncertainty Ratio and Test Accuracy Ratio for Standard M&TE equipment used in sterilization calibrations is governed by individual specifications and plant SOP's and meets the required TAR/TUR of 4:1 or better.

The calibrations department will manage their calibrations generated by computer enterprise planning system. The chemistry, micro, and environmental labs are responsible for their department calibrations. Their calibrations are completed on paper, and those departments enter a pass/fail status in computer enterprise planning system. Other departments that have not been trained on computer enterprise planning system calibrations will return their completed paper calibrations to the calibrations department, who will enter the pass/fail results into computer enterprise planning system.

Note: Assets will also be inspected for physical damage during calibration and production operation. If physical damage is noted during calibration or manufacturing operation, manufacturing supervisor, quality, and calibration will be notified to determine the appropriate corrective action. If visual potential microbial growth is noted, micro lab will also be notified, and the appropriate documentation and an investigation will be initiated including all testing needed.

Whenever the asset status is being changed to decommissioned or inactive, a verification must be performed prior to removal from service. If an asset is made inactive or decommissioned following an acceptable final calibration, computer enterprise planning system calibration records will be updated to reflect the status change. When the calibration department identifies an asset as non-operational, missing, or is notified of a change of asset status, a non-conformance form will be initiated to complete the status change and the updating of the computer enterprise planning system calibration records. An assessment will be performed to determine if an OOT report will be issued if the asset is damaged to the extent of being non-operational.

When it is necessary to change the status of a calibrated asset to inactive or decommissioned, all associated computer enterprise planning system records, i.e., PM record, and job plan record must reflect the change and become inactive. The asset assigned to the data sheet record must be deleted if the asset is decommissioned. Conversely, when returning an asset to service all associated records, i.e., PM record and job plan record will reflect the status change and become active.

The asset will be added to the data sheet record under the asset tab. When a calibration Standard (M&TE) is being removed from service for recertification, the asset status should be changed to reflect: pending vendor service and the PM record inactive. When returning the M&TE to active status the PM record will be returned to active.

Standard M&TEs to be decommissioned must be verified prior to decommissioning if they have been used to perform any calibration since the Standard M&TE's last calibration. The status of a Standard M&TE or asset should be changed to inactive if it is being sent out to an approved vendor for re-calibration and is not expected to be returned to service prior to the calibration due date. The computer enterprise planning system calibration work order status will be changed to in progress until the Standard M&TE or asset is received and inspected and the certification has been verified. Then, the calibration work order may be completed and closed.

Equipment owners are responsible for verifying current calibration labels are present. When an asset is removed from service, it will be identified so as not to be used for calibration or equipment verification. Gauges and other monitoring devices that do not require calibration should be labeled as calibration not required. An OOT report will be initiated for inoperative Standard M&TEs that are placed out of service. A record of all assets, gauges, and devices being routinely calibrated will be maintained in computer enterprise planning system for notification when those assets are due for calibration.

The number of calibration points selected for testing is determined by the usage of the instrument under test, the characteristics of the test unit, and the characteristics of available calibration standards. A minimum of three calibration points is recommended unless the UUT cannot be tested at three points due to limitations in accessing the internal measuring device(s) for test point manipulation or the units have fixed calibration points. For example, light intensity, conveyor speeds, and ambient room temperature/humidity are considered real-time fixed-point calibrations that are non-adjustable, and the UUT remains in the area/system being monitored; therefore, multiple test points are not permissible and one point verification checks are used. Calibrated equipment and/or instruments cannot be operated past the calibration due date and must be removed from service/operation.

As found calibration data results will be documented to the same number of significant digits as the tolerance in the appropriate calibration procedure. When comparing as found data to a procedure tolerance, rounding to the specified number of decimal places prior to judging pass/fail is required. When the digit to the right of the significant digit is equal to or greater than 5, round to the next highest number. If the digit is less than 5, the preceding digit is unchanged.

Example: 25.462 = 25 if tolerance is in whole numbers

25.462 = 25.5 if tolerance is in tenths
25.462 = 25.46 if tolerance is in hundredths

Computer enterprise planning system rounds based on the highest resolution value entered. For example, a calibration data sheet can be defined with a tolerance of 2 decimal places, but if readings from a Standard M&TE are obtained with a tolerance of 3 decimal places, computer enterprise planning system will round the data values with a tolerance of 3 decimals when determining if the values are out of tolerance.

Computer enterprise planning system calculates the calibration set point acceptance tolerance from the initial As Found Input value entered. For example, if an instrument's set point check is at 15.00 ± 1.00 and the As Found Input value observed is 15.24, then the acceptance tolerance is calculated to 14.240–16.240. Computer enterprise planning system adds a third significant figure (a zero) to be able to perform the required calculation rounding back to the tolerance of two significant figures. The same applies for instruments with one, three, or more significant figures. This system design ensures the correct rounding to the number of significant figures listed in the calibration tolerance is achieved so that an accurate determination for in/out of tolerance is made.

Individual specifications and standard operating procedures will outline the specific procedure for each asset to be calibrated. Calibration frequencies are established regarding one or more of the following considerations:

- Frequencies stated in specifications
- Manufacturer's recommended recertification frequency
- Stability
- Degree of usage

Calibration standards or M&TE may be recertified by an outside standards laboratory. Calibration M&TEs are normally cleaned and serviced as part of the recertification process. All M&TEs should be stored in their respective container or storage fixture when not in use. All M&TEs should be stored in a laboratory or office environment. Calibration M&TEs should not be stored in manufacturing areas or warehouse. Manufacturer's instructions should be followed for care and handling of equipment. Standard M&TEs are to be used only for calibration and verification. If Standard M&TEs remain in the possession of an outside vendor to perform contract calibrations, a copy of their certifications will be reviewed and maintained by the calibration department.

Standard M&TEs will be reviewed to verify the presence of NIST traceable certifications. If the certification of the standard M&TE has as-found data within NIST tolerance but has a less than 4:1 TUR, justification will be initiated to provide acceptance of the M&TE by recalculating the TUR using process tolerances. If the as-found data is "out of tolerance," an OOT Report will be issued unless one of the following applies:

a. The Standard M&TE has not been used since its last calibration.
b. The Standard M&TE has not been used at the set point(s) that were found to be out of tolerance.
c. The Standard M&TE has not been used at the set point(s) that were found to have a less than 4:1 TUR.
d. The 4:1 TUR is maintained. This is determined by dividing the process tolerance through recalculating the uncertainty of the calibration system by using each system component.
e. The error of the Standard M&TE is not detectable by the under-test asset's resolution.
f. The Standard M&TE was verified to be within tolerance just prior to shipment, then the out of tolerance situation will be attributed to damage occurring during transport.

g. Correction factors in the amount of the Standard M&TE's error were applied to the calibrated assets, and all test points were found within tolerance.
h. The instrument has been calibrated with a different Standard M&TE since the last calibration and found within tolerance.

If for reasons a Standard M&TE does not have a TUR of 4:1 or better, other methods may be used to ensure the adequacy of measurement. Such methods may include:

a. Uncertainty analyses
b. Guard Banding of the UUT

   Example: Subtract the Standard M&TE's tolerance from the UUT process tolerance. This accounts for the Standard M&TE error.

   A temperature controller has an assigned tolerance of ±5°C, and the best available Standard has a tolerance of ±2.5°C. This would only provide a TUR of 2:1. By assigning a new tolerance to the controller of ±2.5°C, the error of the standard is accounted for.
c. Widening of UUT tolerance limits

In some instances (e.g., UV radiometer), the calibration vendor is not capable of providing uncertainties for the Standard M&TE due to the nature of the test equipment and this is adequate provided uncertainty readings are not obtainable (Table 15.1).

Calibration frequencies for standards using specifications where defined manufacturer and external calibration service recommendations, instrument stability, and degree of usage are determined. Calibration standards are sent to external calibration service vendors on or before the defined frequencies listed in the above table, and any instrument deviation is investigated and dispositioned through the Out of Limit Investigation and/or Calibration Addendum process. Dead weight testers can be verified in house to avoid damage incurred during shipment.

Note: If the gauge blocks are defaced, bent, or altered in such a way to affect the integrity of the certification, then they must be returned for NIST re-certification.

Note: If Quartz Control Plate has been subjected to extremes in temperature or has been mishandled: dropped, broken, cracked, etc., plate must be returned to the vendor immediately for calibration check.

Each asset will be assigned and identified with a calibration number at the time it is placed in the calibration system. If possible, the asset ID number will be engraved into the asset or identified by other suitable means such as label tape if engraving is impractical. Additionally, the asset will be tagged with a calibration label to indicate the asset number, the calibration date, asset description and location, the calibration expiration date, and the calibrated range. The calibration technician will verify the calibration label and asset match prior to every calibration. When available, the manufacturer, model, and serial number will be entered into the asset record.

Care and Handling of Measurement and Test Equipment: All measurement and test equipment shall be handled per manufacturer's recommendations or individual SOP's. When equipment is not in use, it should be stored in its respective container or designated storage area or cabinet. Manufacturer's instructions should be

TABLE 15.1
Calibration Schedule

Calibration Standards (M&TE's)	Frequency
AMP meter AC probe	1 year
Barometers	Initially
Bubble-O-meter	Initially
Conductivity cells/conductivity meter	1 year
DC power supply	Initially
Dead weight testers (pneumatic and hydraulic)	1 year
Decade boxes	1 year
Digital current/voltage meter	1 year
Digital force torque indicator	1 year
Digital pressure indicator/gauge	1 year
Digital thermometers	1 year
Gauge blocks/gauge blocks visual inspection	1 year
Gloss meter	1 year
Hygrometer indicator	1 year
Light meter	1 year
Multimeters/process meters	1 year
Particle counters	6 months
Pin/plug gauge	2 years
Polarimeter quartz control plate	Initially
Portable air flow hood	1 year
Portable electronic manometer	1 year
Precision voltage standards (standard voltage)	1 year
Profilometer roughness standard	2 years
PRT monitor	1 year
PRT probes	1 year
Resistors	1 year
Salt solutions for dew point monitor calibrations	Replace after 1 year
Secondary spectrometric STD	1 year
Smoke photometer	6 months
STAGE micrometer (microscope)	Initially
Standard steel rulers	Initially
Stopwatches	6 months
Tachometers	1 year
Thermocouple probe (Omega)	1 year
UV radiometer	6 months
Weights – up to 500 lbs.	1 year
Weights – 500 lbs. and over	1 year
Weights used for analytical balances	1 year

followed for care and handling of test equipment. A thermograph will monitor the temperature and humidity of the laboratory environment. If calibrations are performed onsite, any environmental conditions that affect or may affect the measurement results shall be recorded.

A calibrations overdue report will be generated monthly and sent to management for review. The calibration status of any asset exceeding the due date and its preventative maintenance record will be changed to inactive in the computer enterprise planning calibration system and removed from service. If an asset was in production beyond the assigned calibration due date and not made inactive, a Calibration Addendum will be issued, and the asset immediately calibrated. If the calibration results are acceptable, then justification of no impact will be documented on the Calibration Addendum. If an OOT result is obtained, additionally an Out-of-Tolerance Report will be issued, and a product impact disposition performed per procedure and any required escalations to a Nonconformance Report (NCR) will be addressed in the OOT investigation. If a Calibration Addendum is not required, a comment shall be made on the calibration record and approved by manufacturing and quality management.

A calibrations trend report will be generated monthly. This provides an overview of the calibrations performed and any resulting OOT conditions found by type of instrument. This report is compiled with a minimum of the following computer enterprise planning system-generated information:

- Number of routine calibrations performed
- Unscheduled calibrations performed
- Calibrations performed within due date
- Number of missing and decommissioned assets
- OOT calibrations
- Number of active and operating assets with an inactive PM Record

The calibration lab trend report will be filed in the Document Center after review by management. Computer enterprise planning system calibration records are continually being updated, and information generated for the compilation of the monthly trend reports will be signed and dated and filed in the calibration lab for a rolling year.

Immediately upon awareness of an OOT result, the person performing the calibration will initiate a tracking number and notify the area manufacturing director and quality section manager. If the OOT result may impact product or process, a NCR will be initiated per procedure. In these cases, the NCR must be initiated immediately in case a Field Alert Report (FAR) needs to be submitted to FDA within three days.

If an item of measurement and test equipment (M&TE) is found to be OOT malfunctions or does not conform to specified or assigned requirement, corrective action shall be taken. The OOT investigative report shall include at minimum:

- M&TE ID number
- M&TE description/type
- M&TE owner/department
- M&TE location
- Date of calibrations when OOT was detected
- Measurement data including tolerance limits and maximum deviation or magnitude of detected error
- Actions taken
- Quality impact assessment to any product or process that used the M&TE

The assessment must determine the impact on any product or process that used the M&TE between the date the OOT occurred and the previous calibration.

For assets found to be outside the specified tolerance limit, the rationale listed below will be used to determine if an OOT Report is required. No OOT Report is required if one of the following applies:

a. The asset has not been used since its last calibration.
b. The asset has not been used at the set point(s) that were found to be out of tolerance. Additional set points above or below the operational range must be verified to be within tolerance.
c. The error between the M&TE and the asset under test is not detectable by the asset's resolution.
d. If the asset was verified to be within tolerance just prior to shipment, then the out-of-tolerance situation will be attributed to damage occurring during transport.

The OOT investigation will include a review of the two previous calibrations. If a second out of calibration is found in this review, action will be taken which may include:

- Decrease interval between calibrations
- Broaden the asset tolerance (if the process allows)
- Replace asset with a more suitable one

An assessment will be made by the quality supervisor/delegate for the impact on product or components due to the out-of-limit condition.

Adjustments may be documented on the original calibration record where applicable or a new calibration record referencing an adjustment reading. Instruments capable of being adjusted should be adjusted to as close to nominal as possible. Adjustable instruments are considered instruments in which there is an electrical or mechanical means of adjusting the instrument for the purpose of reducing measurement error.

When calibrations are performed electronically using computer enterprise planning system, the electronic calibration record stored by computer enterprise planning system will serve as the master record of the original data. A completed Work Order Details report can be generated to show the results of the calibration. When calibrations are performed on paper and are not a Pass/Fail or an initial calibration, the paper record is the Master Record of the original data.

The calibration supervisor or quality department manager's approval is required when:

- **The Use of a Correction Factor Was Applied** – Calibration Addendum or Out of Calibration Report will be initiated using a control number issued to the calibration record.
- **The Asset's Due Date Was Exceeded** – When an Asset has been in service for production use and exceeded the calibration due date, a nonconformance or comment shall be issued and referenced on the calibration record.
- **A Frequency Change** – Calibration Addendum or Out of Calibration Report will be initiated using a control number issued to the calibration record.

When an asset is moved to a new location, the calibration label for that asset will have to be reprinted with the new location value. Prior to printing the new label, the asset must be moved within the asset record of computer enterprise planning system to its new location so the correct location value will be displayed on the label.

CONTINGENCY PLAN/DISASTER RECOVERY

The CMMS system (computer enterprise planning system) is a Web-based application that is hosted in client's outsourced data center. If the system is unavailable, a contingency plan is provided to prevent impacting the ability to maintain calibration records. Problems with or changes to the computer enterprise planning system application (configuration changes, reports, workflow, etc.) will be managed per change control procedures. Computer enterprise planning system will be audited annually per procedure. All operations at client's outsourced data center, including backup and recovery of CMMS data, are controlled by IT SOPs.

16 Clean-In-Place (CIP) Systems

Clean-In-Place (CIP) systems normal range of equipment operation are designed specifying equipment that is used to support the production of finished pharmaceuticals in a cGMP environment. An equipment validation plan is developed to outline the planned tasks and expectations for qualification of the equipment. The URS serves as the input to subsequent project risk assessment activities, design objectives, control strategy, and acceptance criteria for testing and qualification of the manufacturing system (Mastrangelo [41]). The qualification activities are planned as follows:

- Stage 1: Design qualification
- Stage 2: Equipment qualification (IQ/OQ/PQ)
- Stage 2: Process performance qualification (PPQ)
- Stage 3: Continued process verification (CPV)
- Cleaning validation (CV)

Each requirement below is assigned an Item ID, description, and category.

The category will be one of the following:

GMP – Any requirement for a function or process that may have an impact on product safety, quality, purity, identity, or efficacy. Requirements categorized as GMP will be qualified.

Safety – Requirement dictated by site, corporate, or other safety-related agency or body, such as OSHA or NFPA. Requirements categorized as safety and impact the control system or any GMP aspect will be qualified (Table 16.1).

The CIP systems is used to clean and sanitize all solution manufacturing equipment including PTS, mix tanks, and STS piping. The CIP system will consist of permanently installed CIP skids, associated piping, and chemical supply containers. Sanitization will be performed by recirculating hot water throughout the system. An ambient-once through WFI rinse will be performed after sanitization. The equipment and process will be controlled by the control automation system. Data collection will be handled by the control system and communicated to the manufacturing execution system (MES) system electronic batch record (EBR).

The PTS and mix tanks will be cleaned through the tank's discharge piping. The STS circuits will be cleaned independently. The general process steps will include

DOI: 10.1201/9781003224716-16

TABLE 16.1
Abbreviations: Process Control

Abbreviation/Term	Definition
FDS	Functional design specification
CIP	Clean-in-place
CNC	Controlled not classified
EBR	Electronic batch record
HDS	Hardware design specification
Controls system	Mix automation
P&ID	Piping and instrumentation diagram
Process air	Compressed air that directly contacts product
PTS	Powder transfer system
SDS	Software design specification
STS	Solution transmission system
WFI	Water system tested to USP WFI (water for injection) criteria

an ambient water rinse, chemical cycle, ambient water rinse, hot water sanitization, and cool down. Cleaning recipes will be developed for each of the cleaning processes.

PTS

Spray devices will be installed. The control system for the PTS will control the cleaning process and will request water or chemical solution from the CIP skids depending on which step is active in the process.

MIX TANK AND DISCHARGE PIPING

Spray devices will be installed in the top and bottom of the mix tanks. Control system will control the cleaning process and will request water or chemical solution from the CIP skids depending on which step is active in the process.

STS CIRCUIT

The control system for the STS will control the cleaning process and will request water or chemical solution from the CIP skids depending on which step is active in the process.

LIFECYCLE REQUIREMENTS

Vendor is expected to provide the following during the design and installation cycle of the project:

- Written communication of any deviations from client specifications.
- Single point of contact for communication.
- Provide an FDS, SDS, and HDS as required.

- Written communication of changes to the design specification.
- Manufacturing/completion project schedule.
- Written communication of delays in the schedule within 1 day of the change.

Product and Process User Requirements

See Table 16.2.

TABLE 16.2
Cleaning Solutions

Name	Description	pH	Concentration
CIP-100	Alkaline detergent	12.4	Up to 1%
CIP-200	Acid detergent	2.0	Up to 1%

Process Quality Requirements

See Table 16.3.

TABLE 16.3
Critical Quality Attributes (CQAs)

Process Step	CQA	Specification	Acceptance
Cleaning	Chemical carryover	12.5 PPB	Cleaning validation
Cleaning	Endotoxin removal	≤ 0.125 EU/mL	Cleaning validation
Cleaning/sanitization	Bioburden removal	≤ 10 CFU/100 mL	Cleaning validation
Cleaning/sanitization	Product carryover	Residual material removed	Cleaning validation

Process Parameter Requirements

See Table 16.4.

TABLE 16.4
Critical Process Parameters (CPPs)

Process Step	CPP	Operating Range	Associated CQA
Cleaning	Rinse time	As developed	Chemical carryover
Cleaning	Supply flow	≥ 5′/s	Bioburden/Endotoxin
Sanitization	Temperature	≥ 80°C	Bioburden removal
Sanitization	Time	As developed	Bioburden removal
Cleaning	Hot water rinse	As developed	Product carryover
Cleaning	CIP-100 concentration	0.1%–1%	Bioburden removal
Cleaning	CIP-200 concentration	0.1%–1%	Bioburden removal
Cleaning	CIP-100 conductivity	1.227. 11.14 μs/cm	Bioburden removal
Cleaning	CIP-200 conductivity	1.547. 7.614 μs/cm	Bioburden removal

Installation User Requirements

See Tables 16.5–16.10.

TABLE 16.5
Material of Construction Requirements

Description	Category
All product contact steel parts are to be made from 316L stainless steel. Non-product contact may be 304 SS.	GMP
Painted materials are not acceptable. Stainless steel shrouded parts may be allowed if approved by client engineering.	GMP
All welds must be continuous and of the same parent material.	GMP
Welding certifications are required for all product contact welds.	GMP
Acceptable non-metallic product contact materials are Viton, PTFE, Buna-N, EPDM, silicone, and polyurethane.	GMP
Direct product contact parts must have a ≤ 25 micro inch RA finish.	GMP
Porous, fiber-producing, or absorbing materials are not allowed.	GMP
Material certifications are required for all product contact parts.	GMP
Any exceptions to these material requirements must be approved by client engineering.	GMP
All product contact surfaces are required to be passivated. Electropolishing can be done in lieu of chemical passivation.	GMP

TABLE 16.6
Construction Requirements

Description	Category
Distribution system piping shall be appropriately supported and sloped at a minimum of 1/16″ per foot. Slope measurements shall be made between pipe hangers/supports and at each change of direction.	GMP
Product piping and valves shall be constructed using ASME BPE sanitary standards.	GMP
All system low points shall have a low point drain.	GMP
Orbital welded connections are required for product contact piping. All welds will meet local governing SOP requirements.	GMP
All piping shall be clearly labeled with environmentally suitable placards indicating system and direction of flow per GES-UTIL-3020.	GMP
Distribution pumps shall be mounted on base plates to minimize the effect of transient shock waves (water hammer).	GMP
The CIP skid shall be designed as a two-tank (wash and rinse) system.	GMP

(Continued)

TABLE 16.6 (*Continued*)
Construction Requirements

Description	Category
The CIP skid rinses will be directed to drain to the process waste system. Rinses to the process waste system will return to the CIP skid and then be directed to drain.	GMP
CIP supply and return pumps will be centrifugal type and will include automated casing drains.	GMP
The CIP skid shall be capable of delivering WFI, up to two different recirculated wash detergent solutions, and process air for air blow.	GMP
The wash vessel will have the following features: • Insulation • Vortex breaker • Vent filter with heat trace and insulation blanket	GMP
The rinse vessel will have the following features: • Insulation • Vortex breaker • Vent filter with heat trace and insulation blanket	GMP
The heat exchanger will be capable of heating incoming WFI and maintaining the recirculating cleaning solution within the skid boundary at temperatures ≥ 176°F (80°C) prior to leaving the skid boundary and entering the distribution piping.	GMP
CIP chemicals will be supplied from drums or totes using a suction lance and pumped by an air-operated diaphragm pump.	GMP
The flow meter will have the following features: • Minimum flow span: 0 gallons/minute. • Maximum flow span: 180 gallons/minute. • Measurement accuracy: ±0.15% of rate from 10% of span to 100% of span. • Field calibrated span: 0 gpm to 300 gpm • NEMA 4X housing. • ½″ conduit connection. • Continuously adjustable zero and span. • Four-line backlit LCD display with touch control.	GMP
The air-blow assembly will have the following features: • Sanitary 316L ball check valve with automated air-blow system on the discharge side of the CIP heat exchanger. • The air-blow supply filter shall be a sanitary code-7 style with a 316L SS housing and 0.2 μm PTFE filter cartridge. • Vendor shall include a filter/regulator and pressure gauge for the air-blow compressed air supply.	GMP
All piping will be fully drainable.	GMP
The system will be designed to prevent cross contamination of unclean, clean in-process chemicals, and cleaned equipment sections.	GMP

TABLE 16.7
Lubrication Requirements

Description	Category
All lubricants used must be detailed in the design documentation. Include lubricants used in the mechanical aspects of the equipment as well as the lubricants used in the product assembly.	GMP
Where direct lubrication is required, the design and construction must be such that the lubrication cannot leak or drain into the product or into the product contact areas.	GMP
Where pneumatic machine functions require the use of lubricated compressed air, only USP and FDA-approved lubricants are acceptable: e.g., White Mineral Oil, USP.	GMP
Lubricants with USDA and FDA approval and "AA" classification shall be used where incidental contact may occur.	GMP

TABLE 16.8
Pneumatic Requirements

Description	Category
A single compressed air supply must be utilized for both oiled and filtered air. Process air must not be lubricated.	GMP
Process air piping must be made from 316L stainless steel. Tubing must be a non-sloughing plastic material.	GMP
Air lines are to be color coded as follows:	GMP
Yellow – Pressurized air lines for non-product contact air.	
Clear – Process air lines.	
Blue – Vacuum lines.	
Black – Exhaust lines.	
All pneumatic lines must be accurately labeled at both termination points.	General
All pneumatic manifold assemblies must have quick disconnects.	General
All pneumatic components must be mounted within the machine base or enclosure.	GMP
Use venturis for vacuum where possible. Vacuum pumps must be approved by Client Engineering.	General
Air blow-offs are not acceptable in clean room installations.	GMP
A 0.22-µm "point of use" filter will be installed by client, where practical, between the last air control device and product contact air supplies.	GMP
Maximum compressed air pressure available is 90 psi.	GMP

TABLE 16.9
Electrical Requirements

Description	Category
Electrical systems must be designed to the provided site electrical and control specification.	GMP

TABLE 16.10
Safety Requirements

Description	Category
All pinch points and hazardous areas must be guarded per OSHA standards using required materials of construction.	General
E-stop buttons must be located near operator work areas and must release all potential energy immediately upon engagement.	GMP
All access doors must have interlocks as described in the site electrical and control specification.	GMP
The design must permit the electrical, pneumatic, and hydraulic devices to be locked out during maintenance and repair activities. Lock-out points must be easily accessible.	General
The control system must be designed to operate in a fail-safe manner and meet OSHA safety requirements. The safety circuit must be category 3 or higher.	General
Jog capability is acceptable only with a single momentary switch.	General
Sharp corners and edges must be eliminated.	General
Safety circuits must not be able to be bypassed.	GMP
Access doors must be interlocked with magnetic mechanisms to not allow entry to the machine while it is running.	GMP
Electrical panels with >50V are to be designed with incident energy below 1.2 cal/cm².	Safety
Fixed mount HMIs are to be set at a height of 58″ from the floor to the center of the screen. Adjustable mounts are to cover the range of 54″–68″ to center of the screen.	Safety
When an emergency stop button is pushed, all sequences shall stop, and valves shall return to their safe position. All stored energy must be immediately dissipated. Operator reset is required to resume operation.	Safety
Noise level as measured at a 3′ perimeter of the equipment shall not exceed 80 dBA.	Safety
Enclosed spaces must be illuminated to a minimum of 100 Lux.	Safety
Hot surfaces must be insulated or labeled if insulation is not possible to help protect personnel from harm.	Safety
Chemical containers must be labeled with contents and SDS information.	Safety

Operational Requirements

See Table 16.11.

TABLE 16.11
Process Constraints and Limitations

Description	Category
STS piping diameters will range from 1″ to 3″ depending on the line.	General
Available utilities include: • WFI: 150 GPM • Steam: 55 psig • Power: 480V 3 ph 60Hz • Compressed air: 90 psig	General

FUNCTIONS (FCT)

Description	Category
The control system will include recipes with cleaning cycle parameters.	GMP

PROCESS CONTROL SYSTEM

Description	Category
The CIP skid shall be controlled by the Client North Mix Automation system and be capable of providing solution as requested by the system.	General

EQUIPMENT ALARMS AND WARNINGS

Description	Category
Critical alarms shall automatically stop the equipment and notify the operator of the condition(s).	GMP
The operator will be required to acknowledge the alarm before the alarm can be reset and the system restarted.	GMP
The condition leading to the alarm must be resolved before the alarm or warning can be reset.	GMP
Warnings (informational) will notify the operator of a condition but not stop the equipment or require further action.	GMP
Discrete devices valves, actuator position must be monitored in both directions open and closed. The system must fault if a position error is detected.	GMP
Any CQA that is out of specification limits must trigger an alarm.	GMP
System compressed air pressure outside of limits must trigger an alarm.	GMP

DATA

Description	Category
Data collection will be defined in the control system project.	General

POWER LOSS AND RECOVERY

Description	Category
On power restoration, the system shall not restart without operator interface or communication-link input.	GMP
No damage to machine will occur as a result of going to the safe state.	General
The equipment manufacturer will supply instructions for recovery from catastrophic control system failures. It is generally accepted that the system shall protect in the following priority: personnel, equipment, and then product.	GMP

(Continued)

The process may be able to continue operation upon loss of power communication with the server. The ability to continue will depend on where the process is stopped. In some cases, the process will be put into a hold state and resume once the power is restored or communication is re-established. In some cases, the process may have to be aborted. Data stored during the outage will be uploaded once communication is restored.	GMP

CLEANING AND SANITIZING

Description	Category
The CIP skid will be cleaned and sanitized on a regular basis.	GMP
Cleaning validation will be performed on the CIP skid.	GMP

MATERIAL/WASTE MOVEMENT REQUIREMENTS

Description	Category
Diluted CIP chemicals may be discharged to drain if concentration is $\leq 1\%$.	General

MAINTENANCE REQUIREMENTS

Description	Category
Preventative maintenance procedures will be available at the FAT	GMP
Spare parts lists will be available one month prior to FAT.	General
OEM part numbers are required for all spare parts.	General
AutoCAD drawings will be supplied for all machined parts.	General
The Operations Manual will include a section on troubleshooting and repair.	General

TRAINING AND DOCUMENTATION REQUIREMENTS

Description	Category
Material and weld certifications for direct product contact parts are required.	GMP
Material cut sheets or specs are required for all process air connections.	GMP
Training documents will be supplied for operation and maintenance procedures.	General
Operator training will be available during commissioning, potentially on multiple shifts.	General
Maintenance-specific training will be available during commissioning, potentially on multiple shifts.	General
A draft operation manual will be available one month prior to the FAT and the final version at the FAT.	General
Manuals will be provided in both electronic and hard copy versions.	General

17 Cleaning Validation

Based on FDA audits regarding establishing the basis for scientific approach, besides coverage and recovery studies, a matrix study based on DOE (Hi-Lo, crevices, difficult to reach internal structures, tiers, nozzle connections, etc.) to establish justifiable critical swabbing points (sampling locations) is recommended to ensure equipment repeatable cleanliness (batch-to-batch, between batches). This is usually covered under HACCP (Hazard Analysis Critical Control Points) Risk assessment program, which covers cleaning validation.

The validation of cleaning processes for process equipment in a dedicated and/or multi-use environment, both manual and automated, for indirect and direct product contact surfaces applies to entities, functions, and personnel that perform or support cleaning validation activities (LeBlanc [1–4]). This does not apply to product decontamination processes for electro-mechanical devices and/or non-product contact surfaces (Table 17.1).

Each facility shall establish procedures that describe the design and operation of its cleaning validation program as outlined in this procedure (Griffin, Reber [44]). Local procedures shall include the types of equipment (e.g., non-dedicated or dedicated), cleaning agents, and cleaning processes (e.g., manual, or automated) validated and maintained at the facility. The number of runs performed within the validation shall be based on knowledge (process understanding) as well as the overall risk to product, including process monitoring. The rationale to justify the number of runs must be documented. The rationale can be based, but not limited, on the following: development studies, the equipment under validation, risk controls in place, and relative product risk. Minimally, cleaning procedures shall be validated for all product contact equipment that is used to produce one or more commercial or validation lots. This applies to dedicated and non-dedicated equipment. The "test until clean" method shall not be used. Where appropriate, cleaning verification without validation is allowable but shall be justified and documented. This may be appropriate in situations where validation is in process, during refinement of a cleaning procedure, or for clinical material other than validation lots. If a product is not produced frequently, it is expected that validation may take a longer timeframe to complete.

Until such time as there are enough runs completed, then cleaning verification shall be performed following every production lot. Analytical methods used to verify equipment cleanliness and release of equipment shall be validated. In some cases, it may be that the method is developed and tested but not validated. Use of not validated method shall be justified. Where accessible, equipment shall be visually examined for cleanliness following each cleaning validation run. Cleaning validations must be performed on each product/code with either a matrix approach or risk assessment. For equipment with a defined lifespan (e.g., chromatography columns used in manufacturing operations, filter cassettes), cleaning validations shall be performed throughout the equipment use lifespan.

DOI: 10.1201/9781003224716-17

TABLE 17.1
Cleaning Validation Definitions

Term	Definitions
Acceptable daily intake (ADI)	A measure of the amount of a specific substance that can be consumed daily over a lifetime without an appreciable health risk.
Chemical constituent	A substance or ingredient found in the product.
Clean hold time (CHT)	The maximum time that clean equipment can be held without requiring additional re-cleaning prior to use.
Continued process verification (CPV)	Assuring that during routine production the process remains in a state of control.
Coupon	A tab representative of the surface material being cleaned is used for surface testing.
Critical quality attributes (CQA)	Are chemical, physical, biological, and microbiological attributes and/or variables that can be defined, measured, and continually monitored to ensure final product outputs remain within acceptable quality limits.
Critical process parameters (CPP)	Key variables affecting the production process and or attributes that are monitored to detect deviations in standardized production operations and product output quality or changes in CQA.
Dedicated equipment	Equipment that is used to process a single product.
Dirty hold time (DHT)	The maximum time that dirty equipment can be held following production and prior to cleaning.
Hard to clean areas	Areas identified as more difficult to be cleaned
Limulus amebocyte lysate (LAL)	Is an aqueous extract of blood cells (amebocytes) from the horseshoe crab, *Limulus polyphemus*. LAL reacts with bacterial endotoxin or lipopolysaccharide, which is a membrane component of Gram-negative bacteria. This is used in test to quantify endotoxins.
Non-dedicated equipment	Equipment that is used to process multiple products.

Non-dedicated equipment shall be cleaned between the production of different products to prevent cross-contamination. For non-dedicated equipment being cleaned to allow the production of different products, acceptable limits are based on the most conservative of the dose and/or an appropriate calculation for the circumstance (e.g., weight percent calculation). Health-based exposure limits (e.g., ADI) should be used when available.

Where dedicated equipment or campaign production of successive batches of the product is used, equipment shall be cleaned at appropriate intervals to prevent build-up and carry-over of contaminants (Ljungqvist, Reinmueller [46]); intervals depend on microbial growth and/or product degradation. The maximum time and/or the number of cycles between cleanings shall be established and validated.

Note: It may not be necessary to remove residue to the same level as in a product changeover, since carry-over may not represent adulteration of the subsequent lot. When a product campaign includes identical formulations of different potencies, testing shall be performed when sequencing from a higher to a lower concentration to ensure that the quantity of residue of the higher potency formulation will not significantly impact the potency of the subsequent formulation unless the cleaning sequence is validated.

NEW PRODUCTS AND PRODUCT CHANGES

Any new products introduced to a manufacturing facility shall be evaluated with respect to cleaning validation (Prince [50]). Any new chemical ingredient or a change in the concentration of a previously validated component or ingredient shall be evaluated with respect to cleaning validation.

Refer to Table 17.2 for guidance on when cleaning validation and/or cleaning verification runs shall be performed.

CLEANING PROCESSES AND CHANGES

All newly developed cleaning processes require validation. Any changes to existing cleaning products, process, or equipment require a thorough documented evaluation with respect to cleaning validation (Hall [54]). Any changes determined to potentially alter the cleanability per worst-case conditions require a cleaning validation to be performed. Any changes that are determined to fall within the existing worst-case conditions require a documented justification but may not require formal cleaning validation.

RISK ASSESSMENT/MATRIX APPROACH

For both new and revised cleaning methods, a documented risk assessment shall be performed. Risk assessments are an important element of risk management and need to be included early in the process of cleaning development and validation. This process is best driven by a team of subject matter experts that have multiple

TABLE 17.2
Cleaning Validation Decision Table for Products

If It Is Determined That...	Then...
A new product contains a chemical or biological component that represents a new worst-case condition to an existing product group...	Cleaning validation is required per the requirements in this procedure.
A new product falls within the validated existing worst-case condition(s) ...	A formal cleaning validation is not required. However: • The rationale and justification shall be documented. • Cleaning verification confirmatory run or runs may be performed.
An existing product or production process has a change per criteria identified in the risk assessment/matrix approach and worst-case identification sections	Evaluate if the change constitutes a new worst-case condition. • If so, validate in the new conditions. • If not, document the justification. Cleaning verification confirmatory run or runs may be. performed.

experiences from operations, technical services, engineering, quality, and regulatory affairs. There are multiple tools that can be used to complete the risk assessment, and it is up to the team to decide on which are most appropriate for cleaning design. At minimum, the following shall be considered within the risk assessment:

- Equipment design
- Complexity of the cleaning method
- Frequency of equipment use (production throughput)
- Equipment location in the production process
- Single product as compared to multi-product equipment (e.g., non-dedicated)

Regardless of the tools used, the overall approach is to aid in understanding risks, their root cause(s), and potential mitigations. This activity can also help drive the determination of critical process parameters (CPP) and critical quality attributes (CQA).

MATRIX DEVELOPMENT

A matrix approach should include but not limited to an evaluation of the following

- Product families/categories
- Equipment families/categories
- Cleaning procedure similarities and differences

The rationale for including a product in any grouping shall be properly documented. Each facility may divide its products into product categories/families. Each category/family represents a major division that is based on chemical constituents. Validation is conducted on a representative product which then covers the remaining products in that group.

Each facility may divide its equipment into categories/families. Categories/families may be based on similar characteristics including but not limited to:

- Tank configuration
- Solution transmission system
- Instrumentation
- Jacketing
- The presence or absence or agitators

Groupings may be created for related equipment deemed equivalent for the purposes of cleaning. Validation is conducted on representative equipment which then covers the remaining equipment in that group (a process known as bracketing).

CLEANING PROCESSES (MANUAL AND AUTOMATED)

Cleaning validations shall consider the different cleaning procedures being used at a facility. All cleaning procedures shall be evaluated for similarities/differences such as, but not limited to, the following:

- Surface coverage demonstrating complete wetting of all surfaces
- Detergents/solvents used, including concentration and quantity
- Parameters (e.g., temperatures and flow rates monitored at predefined process locations, time durations)
- Manual processes (Aldrich et al. [89])
- Automated process
- Pattern of load configurations

CRITICAL PROCESS PARAMETERS/CRITICAL QUALITY ATTRIBUTES

Cleaning validation requirements are determined from process parameters based on process knowledge, soiling agents, and equipment cleaning aids (mechanical and chemical). Product and process knowledge and quality attributes should be used to define technologies that best support cleaning activities. Since cleaning validation is another form of process validation with a specialized scope, CPP, and CQA should be defined for the development of cleaning cycles, verified during validation, and monitored to provide assurance cleaning maintains a validated state. Examples of CPP and CQA are as follows:

Critical Process Parameters	Critical Quality Attributes
Process temperature	Visual inspection/limit
Process pressure/flow	Chemical residue limits
Process time	Microbiological residue limits
Cleaning agent concentration	Drain ability/drying
Clean hold time	
Dirty hold time	

CLEANING VALIDATION LIFE CYCLE: CLEANING METHOD DEVELOPMENT

Cleaning methods should be developed prior to validation with an equal focus on a robust cleaning method design and the cleaning validation approach. Knowing that the purpose of pre-and post-cleaning is to remove residual product, soils, cleaning agents, and microbial contaminants to an acceptable level, teams designing and developing manufacturing processes and associated equipment need to consider cleaning method development as important as the manufacturing process itself, with heavy emphasis on process knowledge.

Parameters that characterize the cleaning process such as cleaning agents, temperature, physical-chemical properties, contact time, water rinses and specifics about the equipment such as material, sequence of cleaning steps, pathways, and flow rates should be established prior to validation.

Tools such as Quality by Design (QbD) should be considered to assist in identifying requirements, defining cleaning processes, and optimizing development during validation design. During the design of a cleaning process, it is essential that thorough documentation is created, not only for capturing data and resulting analysis, but equally important is what rationale is used to render decisions. There should be efforts to optimize and continuously improve the cleaning process and be able to adapt to changes in the process.

All cleaning methods developed shall include equipment draining and/or drying as the final step.

STRATEGY FOR PROCESS CONTROLS

Regardless of if the process employs manual cleaning or is solely comprised of automated cleaning steps, there needs to be a strategy for each step in the cleaning process. This should focus on reducing variation and steps to take when manufacturing variation occurs.

In automated systems, critical process parameters should be monitored and set with action/alert alarms when limits are approached. In some cases, advanced strategies such as process analytical technology (PAT) could be used to control process conditions and show cleaning effectiveness has been reached (e.g., inline pH and conductivity, or online TOC).

WORST-CASE IDENTIFICATION: PRODUCT/COMPONENT

Validation activities shall be carried out using the worst-case product (component) as this is used as an indicator of cleaning method robustness. Considerations in the selection of the worst-case residual indicators should include (as applicable):

- Highest concentration seen in a code
- Hardest to clean (physical attributes that make it difficult to remove)
- Pharmacological potency of an indicator/component
- The number of ingredients
- Solubility
- Batch size
- Formulation temperature
- Viscosity
- Toxicity
- Routine microbial content and growth potential

Justification shall be documented in local cleaning validation procedures and/or identified within the specific equipment cleaning validation under performance.

EQUIPMENT

Validation activities shall be carried out using the worst-case equipment. Considerations in the selection of the representative equipment should include (as applicable):

- Areas more difficult to clean
- Equipment material types (stainless steel, plastic, etc.)
- Size or surface area

- Dedicated vs. non-dedicated equipment
- Equipment configuration/geometry
- Product contact time
- Cleaning method

Justification for the worst-case equipment selected shall be documented in local cleaning validation procedures and/or identified within the specific equipment cleaning validation under performance.

VALIDATION TESTS/INSPECTIONS: VISUAL INSPECTION

Where accessible, equipment shall be visually examined for cleanliness following the cleaning validation cycle. It will be considered visually acceptable based on the evident absence of foreign material (product, degradants, soiling agents, etc.), for example, no visual discoloration, clumps, particulates, coating, or film. Visual inspection should include verification of draining and/or drying of equipment.

CHEMICAL TESTING

Rinse and direct surface sampling shall be performed to assess chemicals and analytes. Acceptable chemical assays include, but are not limited to, the following:

- Component specific assays
- Total organic carbon (TOC)
- Total Solids
- pH
- HPLC
- Conductivity
- Resistivity
- Osmolarity
- Ultraviolet scans/reads
- Infrared scans

The selection of a method must be explained in the protocol.

MICROBIOLOGICAL TESTING

Rinse and/or direct surface sampling shall be performed to assess microbial content. Acceptable microbiological assays include, but are not limited to, the following:

- Total microbial count
- Total yeast
- Mold count
- Bioluminescence
- Coliform count

ENDOTOXIN TESTING

For equipment endotoxin, rinse samples should be collected at the completion of the cleaning process. An acceptable endotoxin assay would include, but is not limited to, limulus amebocyte lysate (LAL).

Sampling Methods

There are two general types of sampling that are acceptable, direct surface sampling (swab or coupon method) and indirect sampling (use of rinse solutions). A justification shall be provided if only one sampling method is used. Local procedures shall define the specific sampling methods to be used.

Sampling methods shall be appropriate for the analyte being measured. Analytical method validation and recovery studies shall be established for all worst-case product residues and chemical cleaning agents.

Direct Swab Sampling

The direct swab test can be applied to both water-soluble materials and water-insoluble raw materials (or poorly soluble materials).

For swabbing techniques, a representative area of the equipment is swabbed using a swab or wipe and then tested for recovery of the analyte.

Rinse Sampling

Rinse sampling incorporates the use of a solvent to contact all product contact surfaces of sampled items to quantitatively remove the target residue (if present). Rinse sampling may be performed on the final process rinse or on a separate sampling rinse performed directly after the process rinse is complete.

COUPON TESTING

Coupon testing for residues is limited to water-soluble materials or materials that are made water soluble by converting the material into an appropriate salt form. Coupons representative of the all-surface materials being cleaned shall be used. Coupons are inoculated with the target/equivalent material, placed in predetermined hard to clean locations, recovered at the end of the cleaning process, and quantitatively analyzed.

SAMPLING SITES

Sampling sites shall include worst-case sites and be identified in a local procedure and documented in the established protocol for the equipment being validated.

Sampling sites shall represent all material types in the equipment (e.g., rubber, stainless steel, glass, etc.).

Acceptance Criteria

Acceptance criteria shall be predetermined based on a documented scientific rationale.

Residual Levels

Where inspection is possible, no quantity of residue shall be visible on the equipment or utensil after cleaning procedures are performed. When agents regulated by local agencies have defined residual limits (e.g., Environmental Protection Agency, Occupational Safety and Health Administration), these shall be followed. However, if product and agents do not have predefined limits, residual limits should be determined. The following factors may be used for this determination:

- Acceptable levels shall be predetermined based on scientific rationale not to exceed 0.001 (0.1%) of a dose of any product.
- For any pharmaceutical inactive component (e.g., dextrose, inorganic salts) that has no dosage data, the acceptable daily intake (ADI) should be used to calculate the acceptable carry-over. If the value obtained is greater than 10ppm, use 10ppm as the limit.
- For appropriate applications, at least a three-log reduction of carry-over in protein concentration between batches/ campaigns/or groupings should be demonstrated.
- For appropriate applications, toxicity data may be used to calculate limits (e.g., calculation based on PDE as per EMA).

Residual level calculations may be used as a starting point during the development of new cleaning methods or revised cleaning methods. If during cleaning method development, the preliminary results indicate the method is more capable than the acceptance criteria calculated, then the final acceptance criteria should reflect the capability of the cleaning method developed.

Endotoxin Levels

Where applicable, the endotoxin limit shall be no more than the endotoxin limit of the product and/or equipment in contact with the product.

Microbiological Levels

The bioburden limit shall not be greater than the established lowest bioburden level for the product, product family, and/or equipment in contact with the product.

Cleaning Agents/Sanitizer Validation Studies

All sanitizers used shall be validated.

Cleaning agents shall be chosen based on appropriate criteria such as the following:

- Interaction with other cleaning agents in use
- Facility surface materials to be cleaned
- Equipment to be cleaned including its materials of construction
- Any facility-specific processes or interactions that may affect the effectiveness of the sanitizer

The rationale shall be documented in the sanitizer validation study.

Acceptance criteria for the specific sanitizer being validated shall be justified based on use including adhering to local, national, and regulatory requirements.

At a minimum, sanitizer validation studies shall include the following:

- Validation of the neutralization method for the selected sanitizer shall be performed.
- Testing in vitro including the use of positive and negative controls.
- Testing in situ including the use of negative controls.

Testing shall be performed on representative surfaces within environmental conditions for which the sanitizer will be used.

Effectiveness studies for which bioburden reduction is the primary acceptance criterion shall be performed.

Effectiveness studies for cleaning agents for which the physical removal of gross contamination is the primary acceptance criterion (e.g., detergents used to remove gross debris immediately following use) are not required.

Challenge organisms used in the validation of sanitizers shall be selected based on local flora organisms trending for the area of use. The rationale for the organisms selected shall be documented within the validation.

Dependent on the sanitizing agent under validation (e.g., agent with sporicidal properties, cleaning agents used for general sanitization), the following organisms should be considered when determining challenge organisms:

- Gram-positive coccus
- Gram-positive sporulating rod
- Gram-negative rod
- Mold
- Yeast

Local isolates and reference organisms shall be selected for use with sanitizer validation studies. The facilities should determine the number and type of local isolates to be used based on process or environmental monitoring organism trending data of an appropriate time period.

HOLD TIME DEVELOPMENT

The durations between cleaning, sterilization/sanitization, use of processing equipment/components, and cleaning following use shall be subjected to time limits to control the microbial risks associated with equipment usage. A documented risk assessment shall be completed to aid in establishment of the time limits.

All time limits shall be validated as part of the equipment cleaning validation program. Local procedures shall specify storage requirements, time limitations, and actions to be taken if time limits are exceeded.

Dirty Hold Time

The dirty hold time is required to address the maximum time allowed for chemical residues to remain on equipment (potential dry) and for microbial contamination (potential grow). The maximum time that dirty equipment can be held following production and prior to cleaning shall be validated.

Clean Hold Time

The time that clean equipment can be held without requiring additional recleaning prior to use shall be validated.

Following cleaning, equipment that is to be reused shall be stored in a manner to protect it against microbial contamination. In so much as possible to limit bioburden proliferation, equipment shall be held in a dry or drained state.

Additional Hold Times/Cleaning Frequencies

Dependent on the production process, the following additional timeframes may be considered:

- Maximum campaign length (duration and/or number of batches between cleaning operations)
- Frequency of teardown
- Duration between sterilization, decontamination, or sanitization and use (sterile hold time)

CONTINUOUS PROCESS VERIFICATION

Continuous process verification (CPV; cleaning verification) requirements necessary for maintenance of the validated state of the cleaning process shall be defined in local procedures, CPV plan, or validation document based upon risk assessment. Trending of cleaning cycle performance shall be assessed as part of routine monitoring. Facilities shall have procedures in place identifying the investigation requirements if adverse or anomalous results are identified, but the results are within validated parameters. CPV plan and actions should be managed as per dedicated PPQ. Facilities shall perform an evaluation of cleaning validation process CPV periodically.

FAILURE INVESTIGATIONS

Any cleaning validation or routine monitoring testing failure requires an investigation. The investigation shall be performed and documented as per dedicated PQP.

18 Manufacturing Process Validation

Process validation is the collection and evaluation of data from the process design stage through commercial production, which establishes scientific evidence that the process is capable of consistently delivering quality product (Miller [20,21]).

Process validation involves a series of activities taking place over the lifecycle of the product and process:

- Stage 1 – Process design
- Stage 2 – Process qualification
- Stage 3 – Continued process verification

The objectives of this section are to define requirements for process validation, understand the lifecycle approach to process validation, and describe elements of Stage 1, 2, and 3 of process validation. There are different types of validations, namely critical systems validation, design validation, cleaning validation, sterilization validation, analytical instrument validation, software validation, hardware validation, test method validation, and analytical method validation. Nonetheless, this section is concerned with process validation.

A multidisciplinary team is needed to plan and execute the activities involved in process validation: Quality, engineering, manufacturing, research and development, laboratory personnel for both microbiology, chemistry, environmental monitoring, and sterility assurance for total process qualification. The lifecycle approach to validation involves regulations. Process validation is a requirement of current Good Manufacturing Practices (cGMP)

- Finished pharmaceutical products
    - 21 CFR parts 210 and 211
    - Directive 1572
- Medical devices
    - 21 CFR part 820
    - Directive 745

Stages 1–3 of the validation lifecycle come from FDA process validation guidance on therapeutic products. In addition, EU Annex 15, and Global Harmonization task force (ICH) process validation guidance.

The validation lifecycle consists of three stages:

During stage 1, the commercial manufacturing process is defined based on knowledge gained through development and scale-up activities (process design). During stage 2, the process design is evaluated to determine if the process is capable of

DOI: 10.1201/9781003224716-18

reproducible commercial manufacturing (process qualification). During stage 3, ongoing assurance is gained through routine production that the process remains in a validated state (continued process verification). Continued evaluation and trending during stage 3 can identify potential process improvements. This may require additional process design and qualification activities, which will cause a return to stages 1 or 2.

The validation lifecycle is a process flow that details the specific elements of stage 1, 2, and 3 of process validation.

Stage 1 – Process design, start with product requirements definition. Quality attributes identification and product specification development will follow to complete process description and process flow. In parallel, risk assessment scope in terms of failure modes and effects analysis (pFMEA) tabulation is started. User requirements specifications are drafted to initiate stage 1 – studies for process characterization. A control/automation strategy is contemplated at this time. The output from these steps will lead to design review and design qualification.

Stage 2 – Process qualifications are performed in steps starting with installation qualification (IQ). Once all equipment and systems are checked for appropriate connections, operational qualifications protocols are executed to evaluate process parameters and all variables control limits. Following media fills or products are run to ensure that product characteristics and specifications are reproduced for microbiological and chemistry formulation aspects and checking for product quality. All requirements are tracked in a traceability matrix, and a validation summary report is issued.

Stage 3 – Continued process verification is dictated by a plan that would cover heightened monitoring if applicable or conduct routine monitoring making sure that all necessary investigations are documented, and appropriate action is taken as needed.

To ensure compliance, new product development/introductions, the validation lifecycle will always begin with stage 1 and then progress to stage 2 and stage 3. For legacy products that are in routine manufacture, the validation lifecycle begins in stage 3, as there is already a process that has been designed and validated. When stage 3 data indicates a signal, the validation lifecycle may return to stage 1 and stage 2 activities.

The benefits of the lifecycle approach are

- Safety
  - Patient safety assuring the quality of product
  - Operator safety by reducing workplace hazards
  - Environmental hazards mitigation outside facilities
- Business
  - Increased throughput
  - Reduction in scrap and rework
  - Capital expenditure control
  - Reduced in-process and finished goods testing
  - Reduced business risk

- Quality
  - Increased and consistent quality
  - Reduced rejects
  - Fewer complaints related to process failures
  - Improved compliance to regulations and guidelines

Process validation elements in quality systems cover

- Production and process control
  - Manufacturing process validation: user requirements specifications, process design and qualifications, stage 1 and stage 2 guidance for therapeutics, and requirements for trace matrix.
  - Continued process verification: CPV plan and reaction plan.
- Product development and lifecycle management (PDLM)
  - PDLM inputs: I/O traceability guidance and quality attributes identification
  - PDLM Transfers: In-process and release specifications, transfer guides, quality attributes transfer list/checklist
- Product risk management
  - Risk assessment and reduction: pFMEA

The goal of stage 1 – process design is to design a process suitable for routine commercial manufacturing that can consistently deliver a product that meets its quality attributes. Stage 1 activities include:

- Quality attributes identification/product specifications development
- Process description flow chart
- Risk assessment and process parameters identification
- URS development
- Process characterization
- Controls strategy
- Design qualification review

A quality attribute is a property or characteristic that is bound by an appropriate limit, range, or distribution to ensure desired product quality. A critical quality attribute is a quality attribute whose variation outside the specified tolerance has a significant impact on product quality and patient safety.

Quality attributes are identified from product requirements and transferred to manufacturing through product specifications. Quality attributes identification:

- Defines key design characteristics, which translates product requirements at a meaningful level of detail for transfer to manufacturing
- Establishes a ranking system that can be used to set validation requirements and release criteria
- Provides traceability of quality attributes essential design outputs and product requirements

The quality attribute level of detail is more appropriate for transfer to manufacturing than the requirement level of detail. For example, the hanger hole must withstand the primary container system closure removal forces, for closures that are removed through pulling (requirement). The associated quality attributes are (1) primary film thickness, (2) primary film tensile strength, (3) hanger hole shape, and (4) hanger hole vertical distance to seal.

As the manufacturing process is defined, it must be described in a process description, block diagram, or process flow diagram. The process description/flow should detail each unit operation and should be used to assist in the execution of risk assessments and in the development of the control strategy.

A process description shows process inputs, outputs, yields, in-process tests, and reference all documented related reports. A process flow details all major steps and sub-steps of the process, which must function for successful operation and reflect on any interdependence between steps.

A pFMEA is a risk assessment tool that systematically

- Identifies potential failure modes, their causes, and the effect on product and process
- Identifies process parameters that have the potential to affect each quality attribute
- Documents prevention and detection controls for each failure

A risk assessment should be initiated at the start of process design and updated with knowledge gained throughout stage 1 and throughout the lifecycle

A URS describes the quality and business needs for a process or an individual piece of equipment. Minimally, the URS must identify requirements related to quality attributes and process parameters. Other requirements may include:

1. Build requirements
2. Operational requirements
3. Control system requirements (Moldenhauer – refs. 22–24)
4. Maintenance requirements
5. EH&S requirements
6. Calibration requirements
7. Documentation requirements (Jornitz, Madsen [40])

Process characterization or process development studies are performed to better understand the impact of process parameters on quality attributes. The information gained may be used to define parameter ranges and acceptance criteria for stage 2 activities.

Other stage 1 studies may include factory acceptance testing (FAT) or site acceptance testing (SAT) to evaluate equipment against the requirements in the URS prior to accepting equipment from the supplier.

A process control strategy must be developed based on knowledge and experience gained through stage 1 and through the lifecycle. There are two elements of the control strategy:

1. Specific controls related to each potential failure, which may be documented in a risk assessment or control plan
2. A strategy for compiling and reviewing data from specific controls to ensure maintenance of the validated state, which is documented in a CPV plan

CPV is an element of stage 3; however, the planning for a successful CPV program designs during stage 1. A CPV plan does not need to be finalized until the start of routine manufacture.

The proposed process design against the URS must be performed and documented as part of design qualification review, which covers:

1. Approved process flow
2. Approved risk assessment
3. Approved URS
4. Approved list of controls

Stage 2 – process qualification

During this stage, the process design is evaluated to determine if it is capable of reproducible commercial manufacturing. Stage 2 activities include:

- Installation qualification (IQ)
- Operational qualification (OQ)
- Performance qualification (PQ)

Results of IQ, OQ, and PQ are documented in requirement traceability matrix (RTM) and validation summary report. IQ includes demonstrating by objective evidence that all key aspects of the process equipment and ancillary system installation adhere to the manufacturer's recommendati0ons of the supplier of the equipment are suitably considered. IQ verifies that the equipment/systems:

1. Is correctly installed
2. Is connected to specified utilities
3. Is positioned in a suitable environment
4. Has correct supporting documentation

OQ acceptance criteria must be scientifically sound or statistically justified and based on the risk level of the quality attributes documented in the product specifications. Prior to the closure of OQ, the following approval is required:

- Standard operating procedures
- Product specifications
- Cleaning procedures
- Operator training
- Preventive maintenance requirements

PQ includes demonstrating by objective evidence that the process under anticipated conditions consistently produces a product which meets all predetermined requirements, within the scope of validation. PQ must be performed under normal operating conditions:

1. Parameters are within validated ranges
2. Process is in the intended operating environment
3. Operations are using personnel expected to perform these tasks
4. Operating procedures are approved

PQ should occur across a sufficient number of runs (batches) to include any potential source of routine variation such as raw materials, shifts, tank, batch size, or equipment selections. The number of PQ batches must be adequate to provide evidence that the process consistently produces a product which meets all predetermined requirements. Acceptance criteria must be scientifically sound or statistically justified and based on the risk level of the quality attributes documented in the product specifications. PQ must have a higher level of sampling, additional testing, and greater scrutiny of process performance than routine commercial production (Prince [57]).

A separate PQ is not required for each piece of equipment or subprocesses if performed in conjunction with a PQ of the entire process. PQ executed for the entire process meets the intent of the FDA process performance qualification (PPQ) definition for therapeutic products. However, PQ and PPQ may be performed separately as a business decision (Schmitt [69]).

The RTM demonstrates traceability between the requirements in the URS to the protocols and tests where the requirements are challenged to demonstrate test coverage. The RTM must be finalized and approved along with the validation summary report (or included as part of the validation summary report) to ensure that all requirements of validation were met, and the objective evidence is referenced and available (Deeks [86]).

Stage 3 – CPV goal is continual assurance that the process remains in the validated state during commercial manufacture. Stage 3 activities include:

- CPV plan
- Heightened monitoring where applicable
- Routine monitoring and continuous process improvements

By continuing to monitor the process, the impact of equipment fatigue and small changes over time can be assessed. The CPV plan is a tool to document the approach that will be used to collect and analyze data and demonstrate that the process remains in a validated state. It defines the routine monitoring mechanism for validated quality attributes and process parameters relationships. Minimally, the CPV plan must cover all finished good quality attributes.

The initial CPV plan is drafted during stage 1, confirmed during stage 2, and finalized prior to routine production or heightened monitoring. The CPV plan is a living document, and the control strategy must be reviewed and updated throughout the lifecycle as process knowledge is gained. For new products and processes with limited historical data and SME knowledge, continued process verification may include a period of increased frequency of review and sampling following completion of stage 2 (heightened monitoring). If heightened monitoring is completed, it should be documented under a protocol. For new processes, rationale for not completing heightened monitoring must be documented.

Signals during routine monitoring may identify deficiencies in the process or opportunities to improve the process. Product or process changes may require a return to stage 1 and stage 2 of process validation as part of change management. With a robust CPV program, revalidation on regularly defined frequency may not be needed. Revalidation is only needed when the CPV data indicates a return to stage 1 or stage 2 is necessary, or when required by local or international standards.

In conclusion, the validation lifecycle involving process validation is the collection and evaluation of data, from the process design stage through commercial production, which establishes scientific evidence that the process is capable of consistently delivering quality product. Process validation involves a series of activities taking place over the lifecycle of the product and process:

- Stage 1 – Process design
- Stage 2 – Process qualification
- Stage 3 – Continued process verification

19 Risk-Based Life Cycle Management

The risk-based life cycle management (RBLCM) process is a risk-based evaluation of pharmaceutical drugs, medical products (MP), devices, and therapeutic products that ensures adherence to product and regulatory requirements. The process assesses current manufacturing, design, and documentation. In addition, product and process performance data is collected to determine if the current design, manufacturing processes, and risk controls continuously mitigate failure modes that could lead to unacceptable harm to a patient/user or could impact a therapy (Table 19.1).

The playbooks are to be created by cross-functional teams including representatives from plant quality and plant manufacturing, as necessary, including process and product subject matter experts (SMEs). The purpose of the cross-functional team(s) is to conduct a comprehensive analysis for each product family. Under no circumstances shall these deliverables be created by individuals (Table 19.2).

The goal of RBLCM is to:

- Establish objective evidence that demonstrates products and processes are operating in a "state of control." The objective evidence created shall be documented in a product family "playbook" for a single product family or product manufactured or serviced at a unique MP site.
- Periodically analyze the documented risk controls (control strategy) for effectiveness and consistency within and across product families.
- Maintain selected "playbook" deliverables as living documents by continuously enhancing the content as new product/process information is obtained.
- The playbooks serve as mechanisms for knowledge sharing and improvements globally.
- Manage and reduce risk, increasing the overall health of product families by driving action based on ongoing listening system data monitoring and trending, e.g., defect per million (DPM) and complaints per million (CIPM).

The RBLCM process consists of the following main elements:

- Creating Playbook documentation using RBLCM methodology
- Monitoring and trending product performance for each essential requirement over time.
- Analyzing playbooks for adequate risk control within product family and across plants.
- Updating playbook documents related to essential requirements as new information is obtained.

DOI: 10.1201/9781003224716-19

TABLE 19.1
Life Cycle Nomenclature

Terminology	Definition
Control strategy	A planned set of controls, derived from current process and product understanding that assures process performance and product quality.
Critical control point	Point in manufacturing at which controls are applied and data is generated to prevent, eliminate, or reduce the risk related to ERs to an acceptable level.
Essential requirements	Design requirement (characteristic or attribute) of the product that, if not met, can result in harm to the end user. These are provided by the product design owner.
End user	Any internal or external (e.g., patients, caregivers, bystanders, service technicians, environment, etc.) user of a product or a process used to manufacture a product.
Hazardous situation and harm analysis (HSHA)	The HSHA is a systematic tool for the analysis of hazards and hazardous situations specific for a therapy to identify the resulting harm to the end user (e.g., patient/caregiver) and the probability of severity of harm occurring. This assessment is predicated on a therapy and a variety of patient populations. It is also predicated on the types of products used in that therapy.
Listening	
Systems	Internal or external data systems that provide feedback on either product or process performance.
Life cycle	All phases in the life of a product from the initial conception to final decommissioning and disposal.
Other requirements	Non-essential, non-regulatory design requirement that potentially impact product quality and/ or production.
Risk-based life cycle management (RBLCM)	A risk-based evaluation of MP devices and therapeutic products that ensures adherence to product and regulatory requirements.
Playbook	A collection of documents including manufacturing risk analysis (pFMEA), control plan, product performance data, and process and test method validation assessments, created for a finished goods product family or product manufactured at a specific site giving documented evidence that the product and process are in state of control. The playbook deliverables are also used to identify the need for mitigation or improvement activities. Also, a collection of risk-mitigating living documents that will be continually improved within the context of the quality management system as product knowledge evolves.
Product design owner (PDO)	Personnel with the responsibility for issuing product requirements for a specific product family.
Progression level	Defined content and scope of the Playbook. Higher levels are related to increasing product knowledge
Product characteristics	Product characteristics are the measurable attributes of a process step or sub-step. Product characteristics can be found in engineering specifications, drawings, etc. Examples include dimensions, size, tensile strength, etc.

(Continued)

TABLE 19.1 (*Continued*)
Life Cycle Nomenclature

Terminology	Definition
Product family	A grouping of products that have the same or similar intended function/ indicated use, fundamental technology, performance specifications and/ or use in practice, and in general, with exceptions, only differ in non-essential characteristics that do not affect safety and effectiveness.
Quality system improvement plan (QSIP)	A global improvement effort intended to create a new culture that drives continuous improvement through proactive control and mitigation of risk. QSIP is supported by three main pillars, risk-based life cycle management, global product ownership, and quality quotient.
Regulatory requirement	Non-essential design requirements related to a regulatory obligation (e.g. commitment, guidance, standards). If a regulatory requirement is not met, product may be considered adulterated.
Severity	A measure of the possible consequence of a hazard (the consequence of a hazard is the harm to the End User).
State of control	A condition in which the set of controls consistently provides assurance of continued process performance and product quality. Product use test results, complaint data scrap and internal process data are examples.
Therapy	The treatment of physical or mental illness.
Process failure mode and effects analysis (pFMEA)	The pFMEA considers all reasonably foreseeable potential failure modes of each manufacturing process, their causes, and effects on product.
Linkage document	A tool utilized to relate process steps and their associated process parameters to requirements. This tool aids in the creation of the pFMEA.
Process flow diagram	A visual summary of the process steps required to manufacture a product
Control plan	A plan documenting the manufacturing process controls for product and process characteristics.

TABLE 19.2
RBLCM

Member	Responsibilities/Authority
RBLCM core team	Establishing the RBLCM process/procedure
	Approving changes to the RBLCM Process/procedure
	Maintaining alignment of tools and content across product families and manufacturing sites
PDO	Participating in development, update, and approval of essentials

The RBLCM playbook is a summary of objective evidence within the manufacturing space that not only demonstrates legacy product's design and manufacturing processes meet the requirements of the customers, but also demonstrates that products are compliant and safe. This is based on the characterization of the manufacturing processes and the in-process, DPM, and postproduction, CIPM, listening systems. The RBLCM playbooks are created for each finished goods product family manufactured at a specific site.

The "playbook" and other related documents consist of the following elements:

- Product overview (based on product family)
- Essential requirements list
- Process flow diagram
- Process linkage document (historical document)
- Process failure modes and effects analysis (pFMEA)
- Control plan
- Assessment of existing process flow diagram, pFMEA, and control plan
- Assessment of test method validation (TMV)
- Product performance analysis
- Assessment of process validations
- Quality plan
- Gap summary and disposition

All playbooks generated in the site facility will follow the requirements outlined in this document.

- Identify product families
- Identify requirements (ER's) for each product family
- Identify key plant processes that impact the identified requirements
- Complete trace matrix, pFMEA, and control plan for each key plan process impacting identified requirements
- Define internal and external product performance for the identified requirements
- Review validations and test methods for adequacy
- Identify gaps, document in quality plan, act to address, and ensure containment if required

The RBLCM process scope is defined by progression levels (PL):

- **Level 1** – Level 1 is the state before implementing RBLCM process.
- **Level 2** – Regulatory Compliance State: Creation of initial documentation (objective evidence) to demonstrate the state of control for processes and products. Limited use within the life cycle. Level 2 is having completed RBLCM for initially defined essential requirements.
- **Level 3** – Product Performance Improvement: Move from a reactive to a proactive state. Expanding documentation, scope, life-cycle integration, and process analysis to proactively determine product or process issues and improvement opportunities.
- **Level 4** – Continuous Process Monitoring: Continue to progress toward a preventive state: Further expanded scope of manufacturing stream and documentation to create a robust variation reduction management system.
- **Level 5** – Business Ecosystem: Proactive, risk-based system, with holistic integration of quality, manufacturing, and commercial and R&D systems (Figure 19.1).

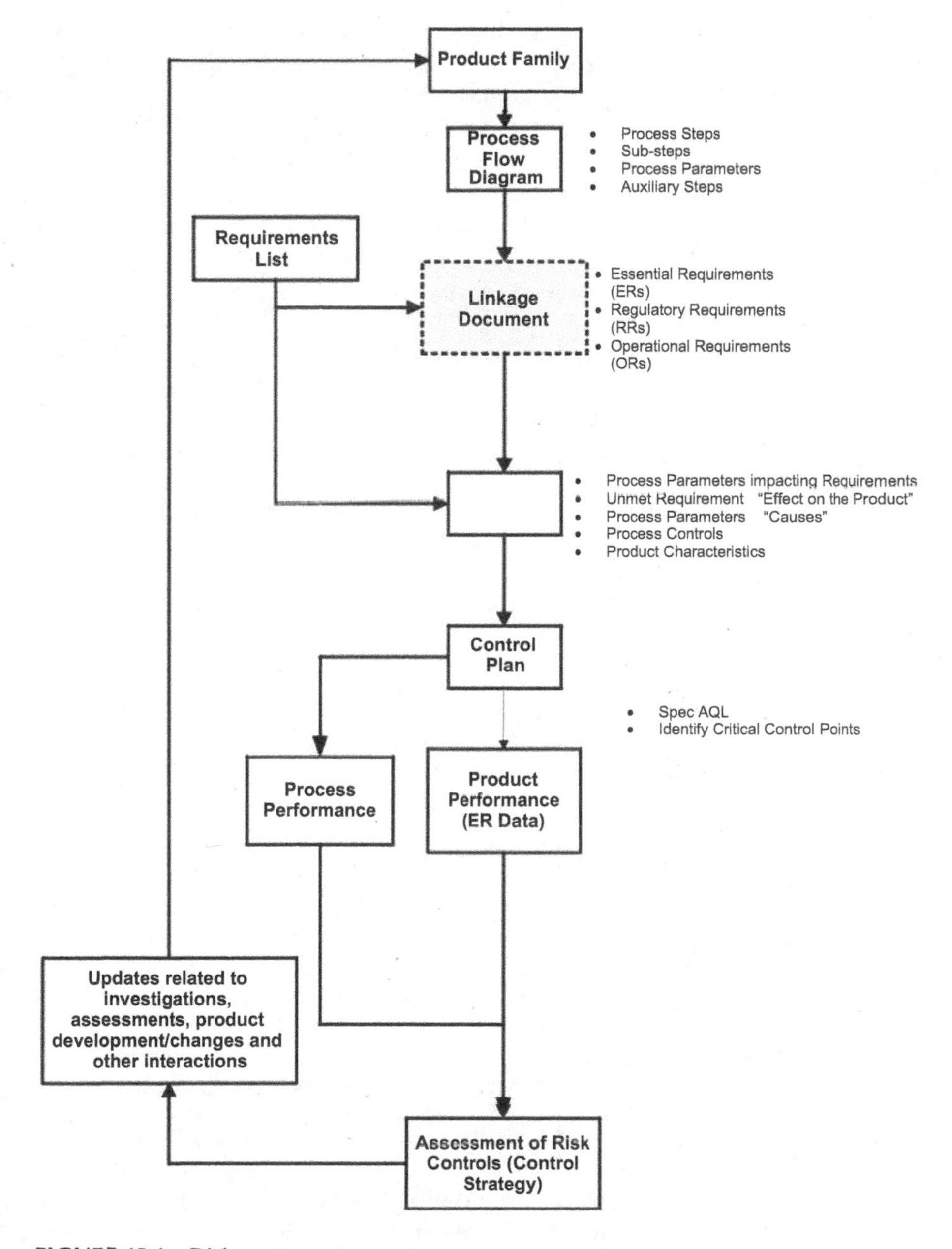

FIGURE 19.1 Risk assessment.

The product family overview document include, at a minimum:

- All applicable product codes manufactured at the site
- A description of the product (include pictures, illustrations, etc.)
- The relevant product history
- A description of how the product is used (intended use)

- Where the product is sold or distributed
- The manufacturing volumes (# units annually) or number of units in field
- Other manufacturing locations for this product family

PROCESS FLOW DIAGRAM

The process flow diagram is a visual summary of the sequence of high-level steps or processes beginning with receiving and inspection and continuing to the production release required for a salable product. Only operations that transform material into subsequent intermediate products to achieve the finished good are considered as process steps. The process flow diagram must include the process step number and name. The step number and name are used to align other RBLCM output documents, e.g., pFMEA, control plan.

ESSENTIAL REQUIREMENTS (REQUIREMENTS LIST)

The product design owner (PDO) provides a set of product requirements (ERs) that will be the foundation of the playbook documents. These describe the ERs by each product family from a patient and caregiver safety perspective.

ERs are the most important aspects of the product family from a safety perspective. ERs are determined by relating critical quality attributes to design requirements/functions. These requirements are then related to hazardous situations from existing risk documentation, such as therapy-specific risk documents including Hazardous Situations and Harms Analysis (HSHA), Clinical Hazards List (CHL), or product-specific risk documents (e.g., Risk Assessment and Control Table (RACT)) to that, if not met, can result in harm to the patient with a severity rating of catastrophic (5), critical (4), or serious (3).

ERs must be defined at a product level (not at the component, process, or subsystem level). The non-achievement of an essential requirement is an end effect on a product and should be relatable to hazardous situations from risk documentation. The PDO will guide and approve the identification of essential requirements.

LINKAGE DOCUMENT (TRACE MATRIX): HISTORICAL DOCUMENT

The Trace Matrix/Linkage Document links the ERs to key steps of the manufacturing process/parameters. This document lists the ERs across the top of the document and process steps and parameters along the side. This document is used to highlight the process parameter relationship and a high-level overview of the process step.

- ERs across top
- Process steps along the side
    - Parameters related to process step function/failure
    - An "x" will identify the process step to the essential requirement potentially affected.

- Every column (requirement) linked to a process step/parameter will be identified with an "x" at the intersection of the column (ER) and row (process parameter/step)
- The linkage document is used to highlight the relationship of the ER to the process. This will assist with the initial pFMEA generation.

The Linkage Document is a tool utilized to relate process steps and their associated process parameters to requirements. This tool aids in the creation of the pFMEA. The matrix documents the relationship between the process steps and their process parameters that may affect an essential requirement. This relationship is determined and noted within the matrix.

Process steps should be aligned between the Linkage Document and pFMEA.

PROCESS FAILURE MODE AND EFFECTS ANALYSIS (PFMEA)

The pFMEA considers all reasonably foreseeable potential failure modes of each manufacturing process, their causes, and effects on product that can affect the outcome of an essential requirement. Based on the failure mode and the causes of the related manufacturing controls, both prevention and detection are documented. At a minimum, the pFMEA must be completed for all process steps and process parameters listed within the Linkage Document that impact Essential Requirements.

The pFMEA identifies the risk associated with the manufacturing steps that can cause the product requirements to be non-conforming. It also identifies the controls that are currently in place to mitigate those risks. The FMEA is to start at the beginning of the process and continue in sequence until the process end. The FMEA will identify the process step, risk (Failure mode), risk identification (controls), risk analysis, and relationship to essential requirement.

- Severity numbers come from the list of essential requirements, given from PDOs. In order to emphasize the overall review process, the severity number for essential requirements has been set to the highest rating of 5.
- Detection rating is assigned based on control methods currently in place. The occurrence rating is based on the estimated defect level that is present considering the preventative measures that are in place.
- The pFMEA is used to identify risk in the process and identify the key controls for the potential failure mode/defect. A risk of 1 is assigned to the optimal condition while a 5 is assigned to items considered to be highest risk, occurrence, or lowest ability to detect. Optimal RPM $1 \times 1 \times 1 = 1$, Worst RPN $5 \times 5 \times 5 = 125$.

The overall rating (RPN) is the Severity × Occurrence × Detection.

CONTROL PLAN

The process control plan (control plan) details the product and process controls from the pFMEA.

The control plan is to document elements of the control strategy utilized at each step to meet the essential requirements and minimize process and product variation. It is complementary to the pFMEA and details the prevention and detection risk reduction measures (controls) identified in the pFMEA. At a minimum, the control plan must include and be completed for all process steps and process parameters listed within the pFMEA. The control plan must include the following information:

- Process step numbering and process name/description consistent with the pFMEA
- Product and process characteristics related to essential requirements per process step
- All in-house processing, inspection, packaging, and release steps that can affect the outcome of an essential requirement
- Equipment for process and testing
- Reference documents for product characteristics
- Validation references for process parameters and test methods
- Characteristics classification for product and process
- Product specifications and process tolerances will be maintained in quality documents summarizing information for the product family
- Inspections
- Sample size, frequency, and person(s) responsible for testing or inspection
- Control method
- Reaction procedure

Assessment of Process Flow Diagram/pFMEA/Control Plan Documents

The assessment is used to evaluate existing process flow/pFMEAs and control plans against an established list of minimum requirements to ensure compliance with the RBLCM process. If gaps are identified within the initial assessment checklist, corrections must be made to the process flow diagram/pFMEA/control plan documents. Upon making the corrections (or creating new document(s)), the assessment will be completed by indicating that all requirements have been met.

Assessment of Test Method Validation Documents

The test method validation assessment evaluates existing test methods for each ER against an established list of minimum requirements to ensure compliance. If gaps to the minimum requirements are identified while reviewing a test method, they must be documented as a gap in the quality plan.

Assessment of Process Validation and Manufacturing Instruction Documents

The process validation and manufacturing instruction assessment were conducted to determine if the processes used in the manufacturing of the product family are correctly validated. This assessment confirms the quality of the validations against a standard provided by the core team. Gaps not corrected will be recorded in the quality plan.

The process validation and manufacturing instruction assessment evaluate existing process validations for each process step identified in the Linkage Document that impacts an ER. These process steps are evaluated against an established list of minimum requirements to ensure compliance. If gaps to the minimum requirements are identified while reviewing a process step, then the gap and mitigation must be noted within the assessment sheet. All gaps not meeting the assessment minimum requirements must be added to the quality plan. This does not include benchmark items.

Product Performance Data

The product performance data details the product performance of the ERs and other important requirements with internal and external measures.

- Data will be reviewed from final inspection test data and complaints.
- The 95% confidence bound will be calculated from the total defects and total samples.
- This upper bound limit will be compared to a 0.065% level and the current AQL defined in the appropriate product specification
- Any item that does not meet the 95% confidence level of the product AQL will be identified as a gap in the quality plan and highlighted.

This data will be updated quarterly. The product performance data is the objective evidence that demonstrates a product family is safe and compliant with current acceptance criteria. Each ER shall have its product performance calculated. The data will be trended after PL 2 completion. The focus is on the collection of data (variable or attribute) from final product (finished goods) testing. If no final product testing is conducted, upstream testing where the requirement was last tested can be utilized. If neither finished good nor in-process testing exists, receiving and inspection testing may be used.

The product performance data report includes:

- Primary Method: product data generated from final product; in-process is compiled over a time period to make a statistical inference regarding how the essential requirements met the specification AQL. Enough data needs to be collected to allow a statistical inference to be made. That is, enough data to obtain 95% confidence limit that the DPM is less than the specification AQL. The defect rate for each essential requirement is calculated for attribute data by using a number of defects over a number of samples and reported as DPM (upper bound 95% defective). Process capability is calculated for variable data and reported as PpK or based on variable capability data and converted to DPM.
- Any failure to meet the current acceptance criteria will be documented as a gap in the quality plan. If an existing CAPA exists for a known problem, the CAPA number will be referenced in the appropriate section.

- Customer complaint data aligned to ERs are used to calculate complaint rate for each ER as the number of complaint instances over released units.
- Create a chart or multiple charts that will show the product performance (% defective/DPM or PpK) over time (for each measurement interval).

Production Process Data Collection (PPDC)

Production process data collection (PPDC) can refer to any production data, including requirement, product characteristic, and process parameter data. As part of PL 3, the PPDC identified shall include the data monitoring of critical control points (CCPs).

CCPs may include ERs, product characteristics, or process parameters. Analyzing data for CCPs as part of PPDC at PL 3 is the primary mechanism to describe process stability for the state of control.

PPDC requirements for PL3 (in progress):

- CCPs shall be referenced in the control plan.
- Data shall be managed per local data management processes.
- Any CCP data collected shall have appropriate control/alert limits defined.
- Control/alert limits shall be within specification limits and derived from process variation as opposed to process capability.
- Manufacturing data sources for CCPs must have objective evidence for TMV.
- Data shall be collected for CCPs at least weekly.
- Rationale for selection of CCPs shall be included in the control strategy assessment.

The quality plan represents a summary of any gaps and the immediate containment action(s) (i.e., additional, or different inspection techniques, increased sample size, restricted ranges of operating values, etc.) that will remain in place until the gaps are addressed. The quality plan captures gaps and their mitigation plans identified during the execution of the process.

- CAPA activity associated with the essential requirements gaps identified
- Defining/addressing gaps
- Trackwise (computerized change control system) nonconformance

The Gap Summary and Disposition Document details a summary of the gap(s) identified within the quality plan during the RBLCM process for a product family. The gap(s) potential impact to customer is/are identified and ranked as either major or minor.

- **Major Gap** – a gap in a product, process, and/or quality system that could potentially result in the manufacture and release of a device or therapeutic that does not meet one or more of its requirements/specifications, and the product defect would have a severity of serious, critical, or catastrophic.

- **Minor Gap** – a gap in a product, process, and/or quality system that could potentially result in the manufacture and release of a product/device or therapeutic drug being that does not meet one or more of its requirements/specifications, and the product defect would have a severity of minor or negligible. Furthermore, this can include a documentation gap that would not affect the ability of released product to meet its requirements/specifications (Table 19.3).

Dispositions 1 or 2 do not require a written rationale and/or action to support the overall assessment and dispositions. In contrast, dispositions 3 or 4 requires a written rationale and/or action to support the overall assessment and dispositions. Rationales for the determination of acceptable or unacceptable risk should be based on the risk management process. Disposition 4 requires a risk-benefit analysis. Based on the information analysis provided and additional pertinent information, a decision is made to determine if the containment and mitigation activities in place provide an acceptable level of risk to the patient based on the benefit of therapy provided.

The Gap Summary and Disposition document shall be reviewed with franchise-specific dispositioning team member(s). The team will review and document the potential impact on patient and categorize the gap(s) identified for a product family as major or minor.

The regional business quality leader and cross-functional team member(s) document rationale for major gaps with a disposition of 3 or 4, if applicable, and sign the Gap Summary and Disposition document.

Assessment of Risk Controls (Control Strategy)

The RBLCM process documentation requires a periodic assessment. At a minimum, this assessment shall be completed once the other requirements have been completed, and annually thereafter. The assessment shall evaluate the failure modes/causes and their related process controls (prevention and detection) along with the product performance and process data of the requirements, product characteristics, and process parameters to evaluate effectiveness. This analysis will provide a foundation to compare controls across product families, plants, equipment, suppliers, service centers, etc.

TABLE 19.3
Gap Summary Disposition Rating

Rating	Disposition Description
1	No gaps have been identified. No product impacts.
2	Only minor gaps identified. No product safety impact.
3	Major gaps identified; however, benefits outweigh the risk of the gaps identified (rationale required).
4	Major gaps identified; full risk-benefit analysis required to continue releasing product.

Maintaining "Living Documents" Updates to the Playbook Documentation

The playbook contains information that documents product requirements, product or process failures, causes of process failures, preventive and detection controls, and product performance among other information associated with a product family and its manufacturing process. Therefore, this knowledge can be used for:

- Identify improvement opportunities.
- **Data/Information Monitoring** – to review data, information, and associated trends to determine if an event or change has occurred to product and/or process.
- **Events** – to identify, investigate, and evaluate events.
- **Changes** – to evaluate, implement, and monitor the effectiveness of changes including new products and processes.

The playbook deliverables are to be considered "living" documents, and as such, they shall be updated as new information related to the requirements is obtained. Potential

TABLE 19.4
Potential Sources of Triggers to Update Playbook

General Description of Trigger	Examples	Requirements
Product family GAP	ID new gap Gap closure	Event
New product/ requirement or change to requirements	Code merged/added into product family Product line extension Regulatory requirement Performance requirement Essential requirement	Changes improvements
Proactive period review	Annual product review Risk review Management review Plant reviews	Monitoring
Product family code, volume, description change at plant	Change in codes manufactured at plant associated with exiting product family Change in volume Update to product description	Changes
Nonconformance in equipment, calibration, raw material, or process	AQL failure OOL/OOT Supplier defect Audit observation (internal/external) Equipment set up	Events
Nonconformance of released product	Complaint	Events

sources or triggers that may require a change in the content of the "Playbook" deliverables are summarized (Table 19.4).

Document Management

The initial RBLCM playbook is approved and filed as a protocol document. As part of the progression of the RBLCM process, a quality management SOP system (RBLCM) is used for updating the playbook sections.

The RBLCM playbook will be a living document that will be maintained by the specification department and are viewable online. The product overview, ERs, trace matrix, pFMEA, control plan, and process performance data will be listed.

The documents may be updated based on updated controls, process changes, validation activities, CAPA, complaint investigations, etc.

Updates to the Performance Section are performed by quarter. This update to performance is to provide a larger amount of data for trending. This requires an annual review/audit for current references and compliance. Playbooks sections are approved per the RBLCM.

Playbook Updates/Approvals (Live System)

The process flow diagram, pFMEA, control plan, and process performance updates must be reviewed as required. Changes to the product overview, test method validation, and process validation assessment will be approved by quality engineering, RBLCM representative, and other additional reviewers as needed. Changes to ERs will require approval from the product design owner, plant manager, quality manager, and RBLCM representative.

The playbook will be controlled, which includes storage and change management. The SME (RBLCM representative) and approver will verify the affected areas. The initial playbooks are stored as protocols and are maintained in the Site Doc Center.

RBLCM Data Collection

RBLCM data is collected for monthly trend reports as well as quarterly product performance data trend reports. The following data is collected for trending:

- Product use testing
- Packing defects

Product use testing data and leaker data from the leaker analysis system are also collected. Control records are created to link certain defects to the ERs and to group product codes by families. The exported data is linked to these control records for defects, and complaints can be trended by product family for each essential requirement. Batch production totals are obtained to calculate DPM and CPM. File security is applied to the new data to ensure it is not modified once it is verified (Figures 19.2 and 19.3).

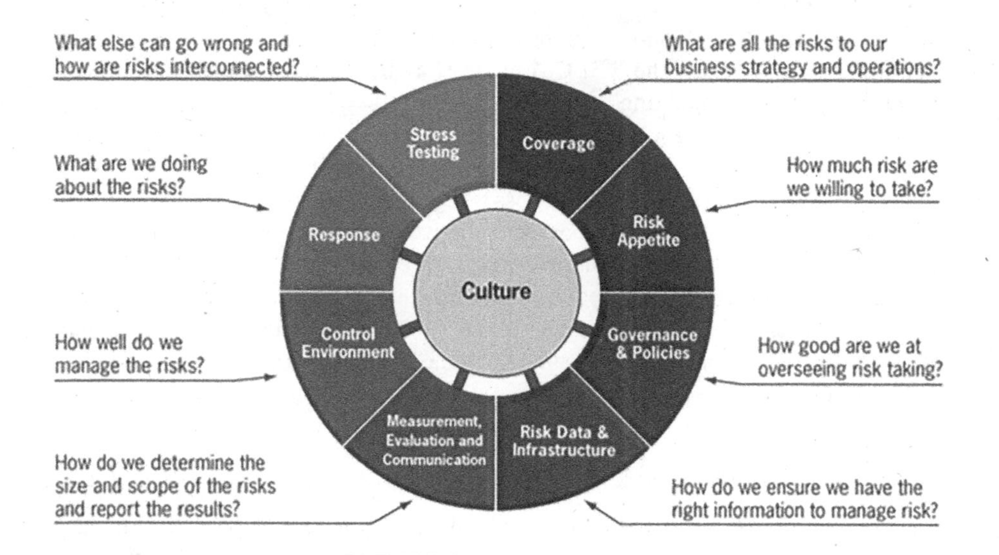

FIGURE 19.2 Enterprise risk management.

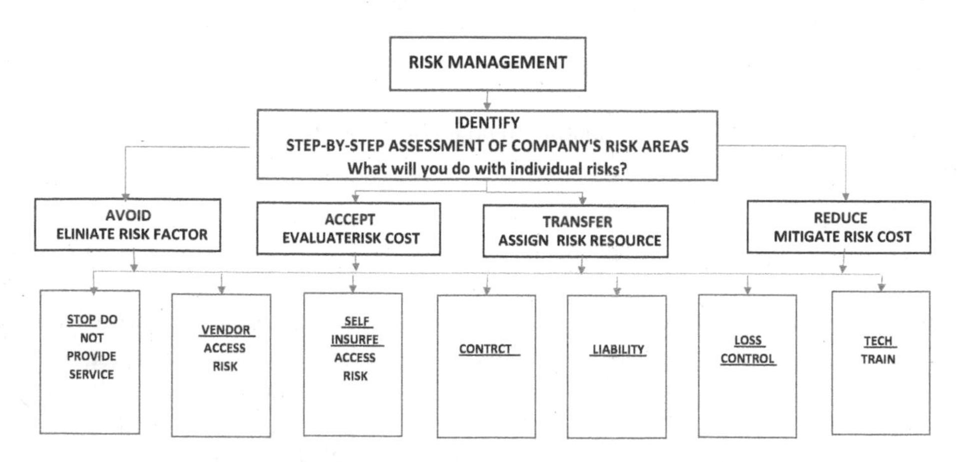

FIGURE 19.3 Process risk management.

20 pFMEA Manufacturing Procedure

This outlines the relationship between the individual process (equipment) Failure Mode and Effects Analysis (FMEA) and the products produced. Each manufacturing area has a table that aligns the process of FMEA required for each product type. This is completed for all manufacturing areas: component preparations, filling, sterilization, and packing. The combination of various areas of FMEA provides the overall risk assessments by product type. This procedure is used in combination with several other site procedures to understand and perform risk assessments. In order to assess manufacturing risks for site manufacturing/production equipment processes, FMEAs are created and documented. These FMEAs for components, filling, and packing production areas are generated per requirements. In addition, overall risk assessment is also completed per procedure. This risk assessment is a start to the end of facility assessment that focuses only on essential requirements. The systems complement each other by allowing independent reviews to be completed.

An overall High-Level Risk Analysis (HLRA) govern an overview of site systems, processes, and tools that comprise the risk management system.

Other requirements related to the site facility validations outline the procedure of FMEAs related to manufacturing equipment and processes. Sterilizer vessel qualification and validation process parameters are covered under specifications, and sterilization process FMEAs.

Site FMEAs are assigned unique control numbers. If an existing FMEA is updated, the previous issue becomes superseded. Change history is documented. Site manufacturing-related FMEA are categorized by product types. These categories are based primarily on product size and represent the different product types manufactured at site. As FMEAs are created or revised, they require quality engineering (or the FMEA initiator) to determine if FMEA needs to be updated. Potential reasons for updates include new equipment, new technology, product/process change, equipment modifications, etc.

All classified environments in which intermediate and final products are manufactured or manipulated, including the storage of components and solutions and the product itself, have been evaluated per environmental FMEAs. Environmental FMEAs are used as part of the overall environmental monitoring program. Environmental FMEAs are controlled per this procedure.

- SITE – Environmental FMEAs are the risk matrix for environmental FMEAs for risk ranking the three risk categories of "Likelihood of Transfer," "Source Condition," and "Detection/Controls." A physical review of the applicable

DOI: 10.1201/9781003224716-20

area is required when updating the FMEAs in order to observe the source condition of the process, room, etc. This rating scale is sensitive to the evaluation of an area for environmental assessments in comparison to the process rating scale where occurrence is measured, severity is based on AQL, and occurrence on a physical test. Environmental FMEAs are risk assessments maintained as living documents and must be reviewed when a change is performed per procedure. Environmental FMEAs must evaluate the opportunities for contamination to the process solution and final filled product container through any openings or apertures during manufacture. Items to consider when creating and updating environmental FMEAs should include the following:

- Facility layout
- Cleanroom equipment, container, and tools
- Process materials (e.g., raw materials, components, container, and closures)
- Product pathway (formulation tanks, solution transmission systems, open versus closed systems):
  - i. Tanks
  - ii. Piping (flexible hoses), air flush, vacuum, or gas systems
  - iii. Agitators
  - iv. Pumps
  - v. Valves/gaskets
  - vi. Change boards
  - vii. Filter housings
- Filter integrity test stations
- Cleaning/sanitization efficacy
- Physical routes of product transfer and associated controls
- Utilities (compressed gases and solutions)
- Available relevant microbial data (e.g., bioburden and endotoxin data from routine monitoring and validation)
- Product-specific data (e.g., growth promotion characteristics)
- Known or anticipated locations of product contact
- Locations likely to have a high potential for microbial occurrence
- Operations involving open or exposed product
- Critical interventions that are performed during the manufacturing process
- Locations where product is exposed to the environment
- Line and process configuration
- Change rooms should be provided between cleanrooms of different classification. These rooms are intended to provide physical separation between areas of differing classification.
- Handwashing should be provided for personnel use prior to entry into change rooms. In lieu of handwashing, local sanitization procedures may be provided with appropriate validations.
- Entry of personnel and items for manufacture use (i.e., equipment, instruments, maintenance equipment, and tools) should be based on procedure to minimize and/or prevent the transfer of microbial hazards into areas of higher classification.

Materials such as paper and cardboard which shed particulates should be avoided. Wooden items (such as wooden pallets) which absorb and harbor microbial risk due to their porous properties shall not be permitted in the classified areas. Consideration shall be given to floors, ceilings, and walls relative to be able to be maintained in a "Dry" state. These include but not limited to coving, material of construction, cleanability, and resistance to agents used for maintaining their use. Procedures should be developed and routinely monitored for spillage response. These procedures should be aligned with the type and magnitude of the spillage to ensure removal to mitigate the risk of contamination/microbial transfer. All classified areas air quality shall be qualified and maintained. Transitions through areas of different classifications shall be controlled to maintain the designated air classification for the activity occurring in the room into the transition area lands. For newly constructed or significantly redesigned facilities and equipment, consideration should be given to the ability for maintenance and/or repairs performed outside the critical areas – outside of Grade A/ISO 5 and Grade B/ISO 7 areas.

21 Analytical Methods Development, Validation, and Transfer

When planning development timeline, the transfer of analytical methods is rarely observed as a major critical item in drug development. Experience has shown based on contract manufacturing operations that many clients misrepresent the level of development on these critical methods. It is significant to assert that validated methods must capture the span of product specification limits to ensure repeatable results that will reflect the validity of manufactured exhibit batches under cGMP conditions. When a method doesn't perform as expected, it can take precious days to investigative work to find the problem, and even more weeks to correct it. To learn about common pitfalls that cause method transfer failures, how to decide what type of method transfer makes sense for a developmental study, and how to implement a fail-safe communication plan that will help put outsourcing relationship on the right course is a critical undertaking in the overall drug development process.

Technical expertise in Chemistry Manufacturing and Controls (CMC) program design, analytical development, and regulatory submissions is of major consideration in pharmaceutical development. Leading CMC development programs for a wide array of biopharmaceuticals, including parenterals, inhalation drugs, and other pharmaceuticals with complex delivery systems, are all important factors in assigning analytical methods. Developing subject matter expertise in HPLC and GC method development and validation, extractables, and leachables program design are key techniques in regulatory submission requirements. Drafting analytical methods in stages of development, validation, and transfer of these critical applications are important in sequencing multiple IND, NDA, and ANDA submissions. Analytical methods are part of submissions to regulatory agencies including responding to FDA deficiency letters. Technology transfer, methods, and techniques transfers such as assay validation of API substances, drug products compositions, residual impurities, elastomeric closures, and seals defects and leakers impacts on product submissions, following stability studies, are all dependent on validated analytical methods.

Analytical biochemistry laboratories are involved in contract research that delivers a broad array of product development and analytical testing services to the pharmaceutical, biotech, animal health, and chemical industries. They support all stages of large and small molecule drug development with expert analytical support, custom synthesis and radiolabeling, and environmental assessments of drug development.

DOI: 10.1201/9781003224716-21

HPLC

High-performance liquid chromatography (HPLC) is a form of column chromatography that pumps a sample mixture or analyte in a solvent (known as the mobile phase) at high pressure through a column with chromatographic packing material (stationary phase). All forms of chromatography work on the same principle. They all have a stationary phase (a solid, or a liquid supported on a solid) and a mobile phase (a liquid or a gas). The mobile phase flows through the stationary phase and carries the components of the mixture with it.

Buffering is commonly needed when analyzing ionizable analytes with reversed-phase LC. For compounds like these, the pH of the mobile phase determines whether they exist in the ionized or non-ionized form. Buffers are also sometimes necessary for applications because impurities or interfering compounds are ionizable. The A solvent is generally HPLC grade water with 0.1% acid. The B solvent is generally an HPLC grade organic solvent such as acetonitrile or methanol with 0.1% acid. The acid is used to improve the chromatographic peak shape and to provide a source of protons in reversed-phase LC/MS. pH changes very little when a small amount of strong acid or base is added to it. Buffer solutions are used as a means of keeping pH at a nearly constant value in a wide variety of chemical applications. For example, the bicarbonate buffering system is used to regulate the pH of blood.

The separation principle of HPLC is based on the distribution of the analyte (sample) between a mobile phase (eluent) and a stationary phase (packing material of the column). Hence, different constituents of a sample are eluted at different times. Thereby, the separation of the sample ingredients is achieved. Retention time (RT) is a measure of the time taken for a solute to pass through a chromatography column. It is calculated as the time from injection to detection. The RT for a compound is not fixed as many factors can influence it even if the same GC and column are used.

The components of a mixture are separated from each other due to their different degrees of interaction with the adsorbent particles. This causes different elution rates for the different components and leads to the separation of the components as they flow out the column.

C18 columns are HPLC columns that use a C18 substance as the stationary phase. C18 simply means that the molecules contain 18 carbon atoms, so the other atoms in the molecule can vary, leading to significantly different substances.

HPLC is a chromatographic technique used to split a mixture of compounds in the fields of analytical chemistry, biochemistry, and industrial chemistry. The main purposes for using HPLC are for identifying, quantifying, and purifying the individual components of the mixture. HPLC is a chromatographic method that is used to separate a mixture of compounds in analytical chemistry and biochemistry to identify, quantify, or purify the individual components of the mixture.

The term reversed phase describes the chromatography mode that is just the opposite of normal phase, namely the use of a polar mobile phase and a non-polar (hydrophobic) stationary phase. Reversed-phase chromatography employs a polar (aqueous) mobile phase (hydrophilic). As a result, hydrophobic molecules in the polar mobile phase tend to adsorb to the hydrophobic stationary phase, and hydrophilic molecules in the mobile phase will pass through the column and are eluted first. Reversed-phase chromatography is the most common HPLC separation technique and is used for separating compounds that have hydrophobic moieties and do not have a dominant polar character (although polarity of a compound does not exclude the use of RP-HPLC). HPLC stationary phases can be segregated by their ability to separate either polar on nonpolar compounds, that is, reversed-phase materials (C18, C8) strongly retain nonpolar solutes with polar solutes eluting at or near the void volume, and hydrophilic interaction. Silica gel is a polar adsorbent. This allows it to preferentially adsorb other polar materials. When it comes to polarity, materials interact more with like materials. This principle is particularly important to many laboratories, which use silica gel as the stationary phase for column chromatography separations.

HPLC: Case Study

Performing analytical methods might involve the determination of product hue to determine if the drug product solution in a cartridge was contaminated by exposure to a colorant. Extraction of colorant might be performed by using methanol and acetone as extraction solvent. The extracts of colorant and drug product solution in contaminated units would be analyzed by using LC/UV/MS screening method. Extract of drug product in clean units would be used as control (Table 21.1).

The instrument used a PDA detector collecting data in the range of 210–400 nm. UV chromatograms were extracted at 225 nm.

For the colorant determination, vialed clear extract solution was analyzed by using LC/UV/MS screening method. For the determination of residue in drug product, drug product was transferred into an HPLC vial and analyzed directly.

0.05 μg/mL of colorant in methanol can be detected in UV channel and observed in extract chromatogram of APCI (+) channel (Figures 21.1–21.12 and Table 21.2).

Analysis Results for Extract in Methanol (Extract Chromatogram)

In conclusion, similarly, samples were tested up to $\lambda = 300$ nm. The analysis result showed that there was no colorant residue observed in product solution samples. No deviations were noted during the study.

Guidelines: Formulating Mobile Phases for Various Reversed Phase HPLC Columns

Different manufacturing processes are used to produce packings for octa-decyl-silyl HPLC columns. The silica base surface area of these columns also differs, according to base coverage with the reversed phase. These factors dictate that different types of columns require different proportions of solvent and water in mobile phases used for the same analysis. These factors also cause variation in the column-to-column performance among columns of any one type.

Methanol-, acetonitrile-, and tetrahydrofuran-water mobile phases needed to elute benzene derivatives are compared. Chromatographers can use these data as guidelines for formulating new mobile phases when they adapt their analyses to highly uniform C-18 columns.

TABLE 21.1
LC/UV/MS Method Conditions

Chromatograph	Type	Ultra-high pressure liquid chromatograph (UHPLC)		
	Model	Waters acquity, agilent, or equivalent		
Column	Type	Acquity UPLC BEH C18, or phenomenex luna C18		
	Dimensions	3 mm × 100 mm, 2.7 μm		
	Temperature	60°C		
	Flow rate	0.7 mL/minute		
Injector	Injection volume	5 μL		
	Loop size	10 μL		
	Injection mode	Partial loop with needle overfills		
Solutions	Mobile phase A	20:80 Methanol/water (v/v)		
	Mobile phase B	Methanol		
	Strong wash	70:20:10 Methanol/water/IPA (v/v)		
	Weak wash	80:20 Water/methanol		
	Seal wash	90:10 Water/methanol		
Gradient program	**Time (minutes)**	**Mobile phase A (%)**	**Mobile phase B (%)**	**Curve**
	0	80	20	NA
	12	5	95	6
	15	5	95	6
	15.5	0	100	6
	18	0	100	6
	19	80	20	6
	21	80	20	6
MS detector	Type	Single quadrupole (SQD)		
	Ionization mode	APCI(−) and APCI(+)		
	Detection mode	Scan		
	Scan range	110–1,400 amu (−); 110–1,400 amu (+)		
UV detector	Type	Photodiode array (PDA)		
	Wavelength	225 nm		

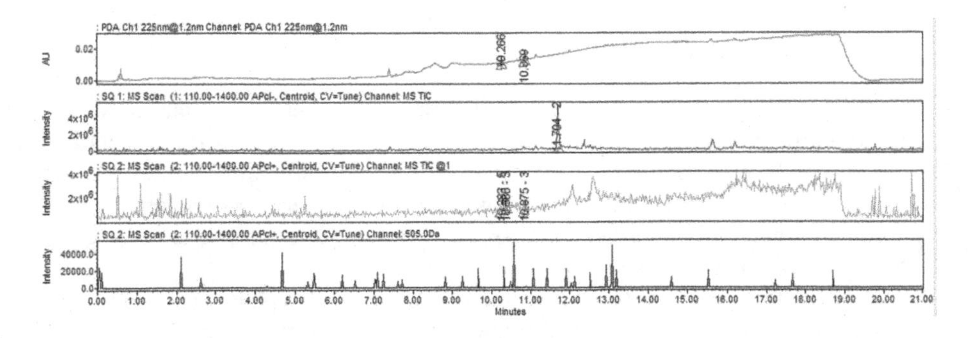

FIGURE 21.1 UPLC/UV/MS traces for clean extract in methanol.

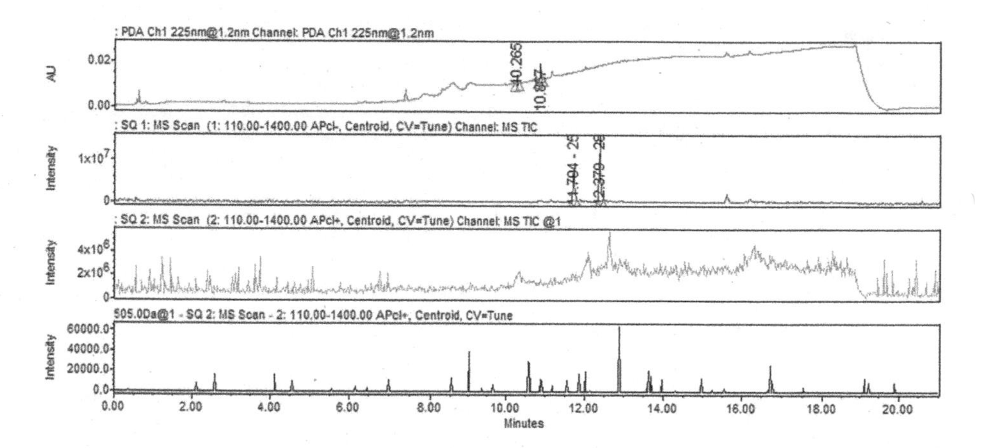

FIGURE 21.2 UPLC/UV/MS traces for colorant in methanol.

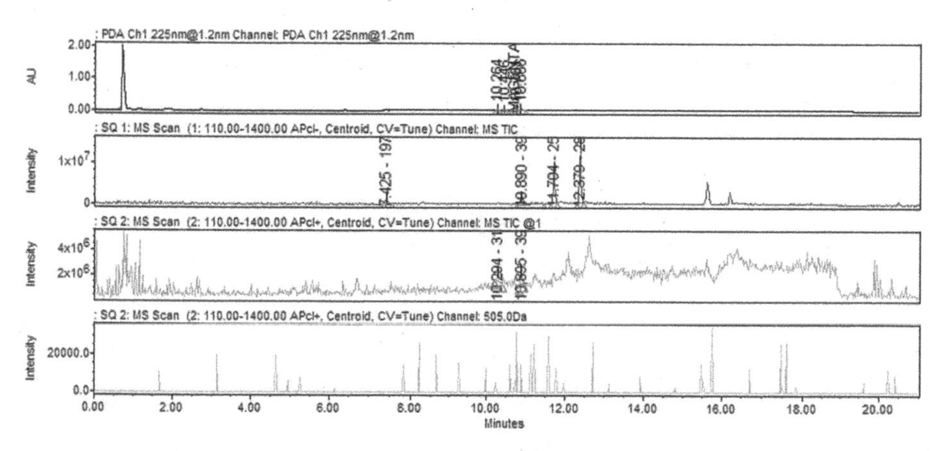

FIGURE 21.3 UPLC/UV/MS traces for clean extract in acetone.

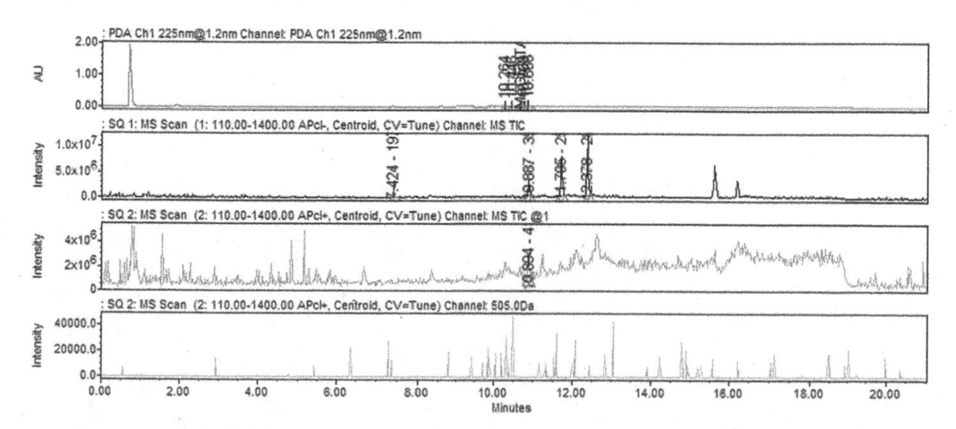

FIGURE 21.4 UPLC/UV/MS traces for colorant in acetone.

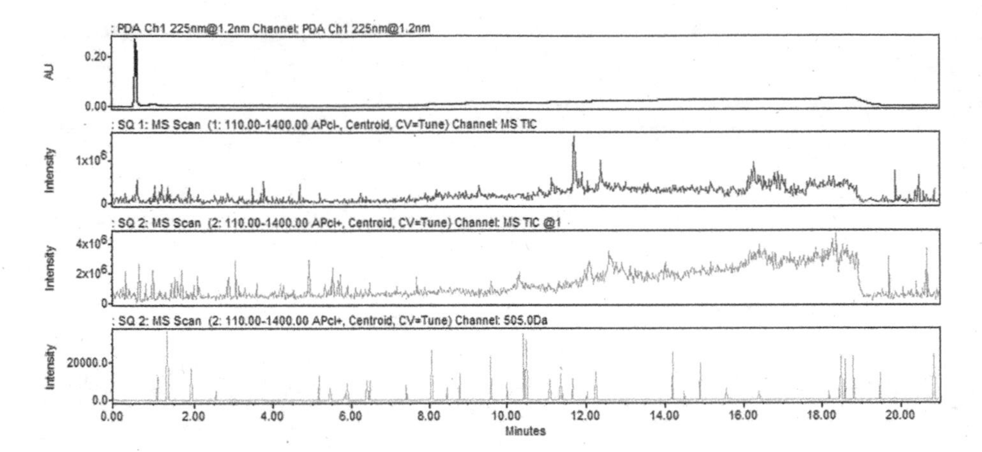

FIGURE 21.5 UPLC/UV/MS traces for drug product – clean unit.

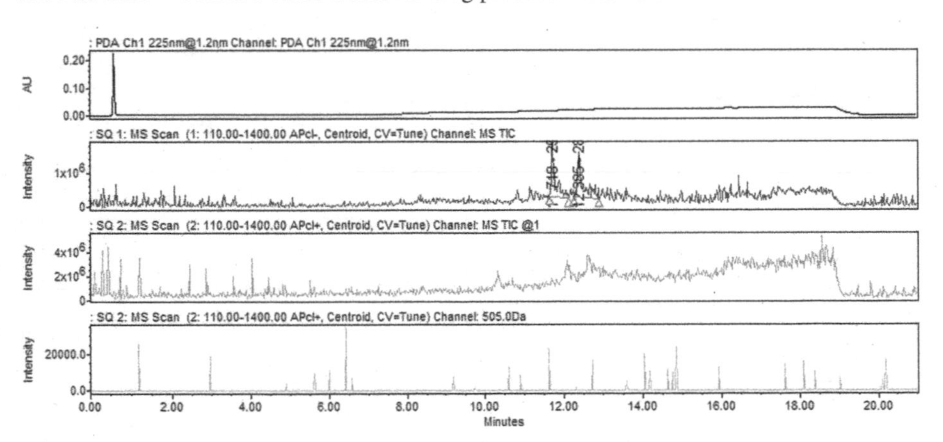

FIGURE 21.6 UPLC/UV/MS traces for drug product – contaminated unit.

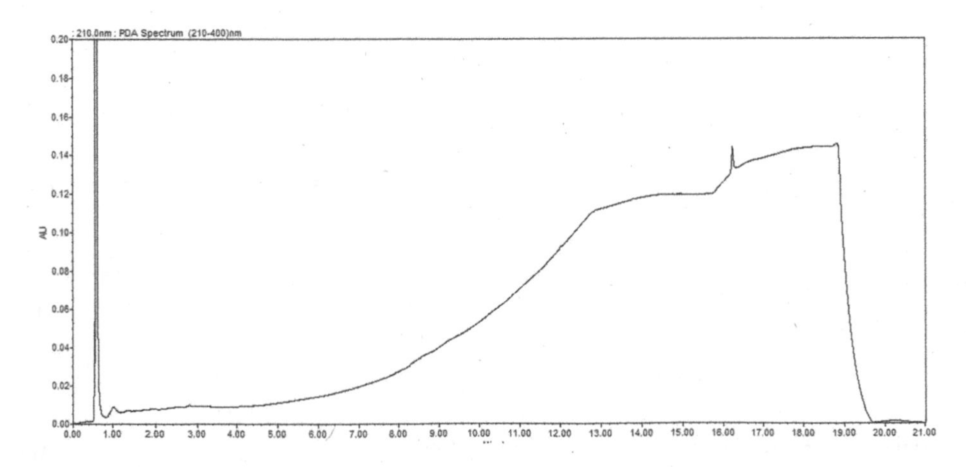

FIGURE 21.7 UPLC/UV/MS traces for drug product – clean unit ($\lambda = 210\,nm$)

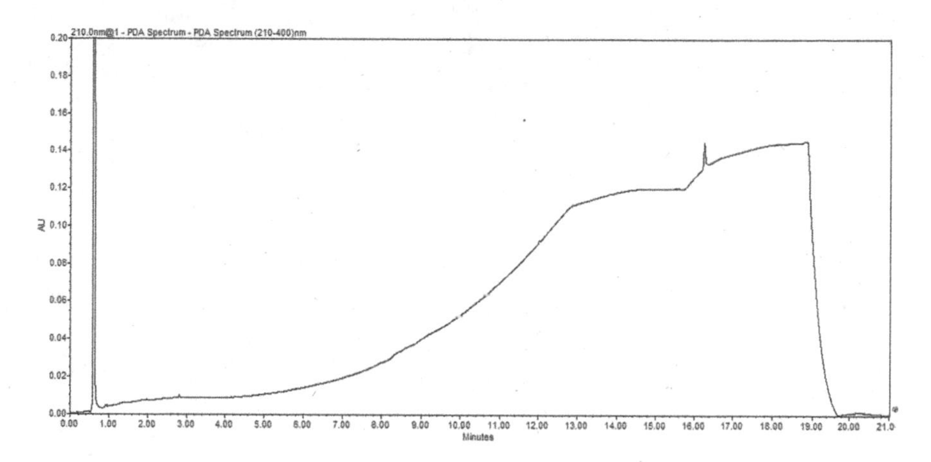

FIGURE 21.8 UPLC/UV/MS traces for drug product – contaminated unit ($\lambda = 210$ nm)

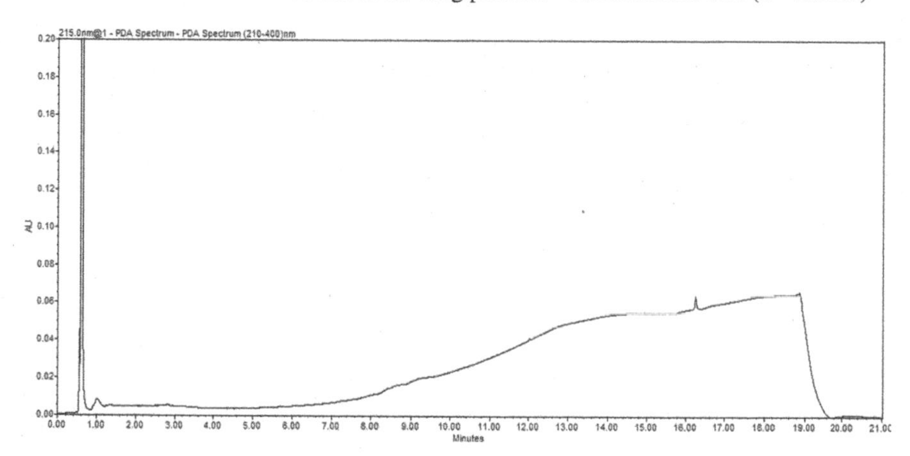

FIGURE 21.9 UPLC/UV/MS traces for drug product – clean unit ($\lambda = 215$ nm).

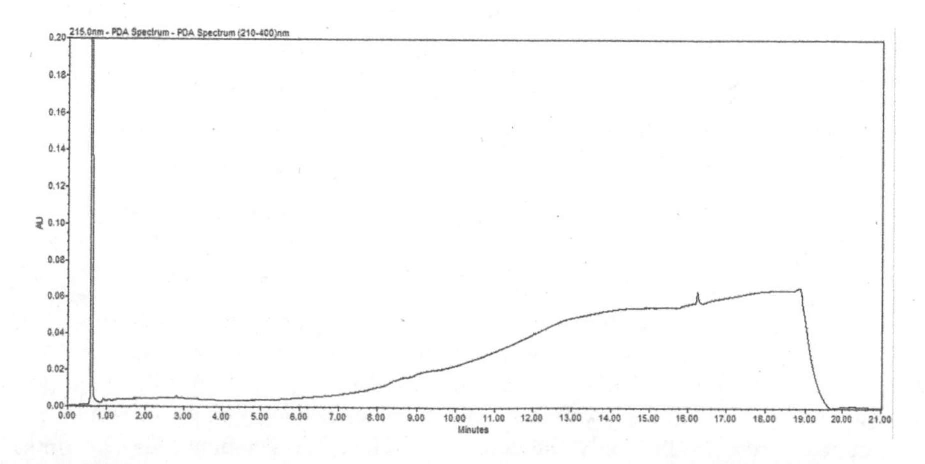

FIGURE 21.10 UPLC/UV/MS traces for drug product – contaminated unit ($\lambda = 215$ nm)

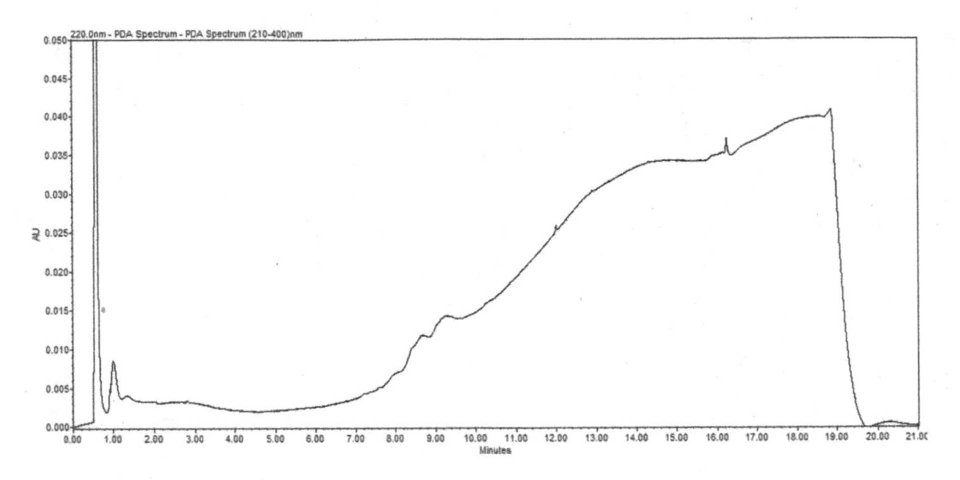

FIGURE 21.11 UPLC/UV/MS traces for drug product – clean unit ($\lambda = 220\,\text{nm}$)

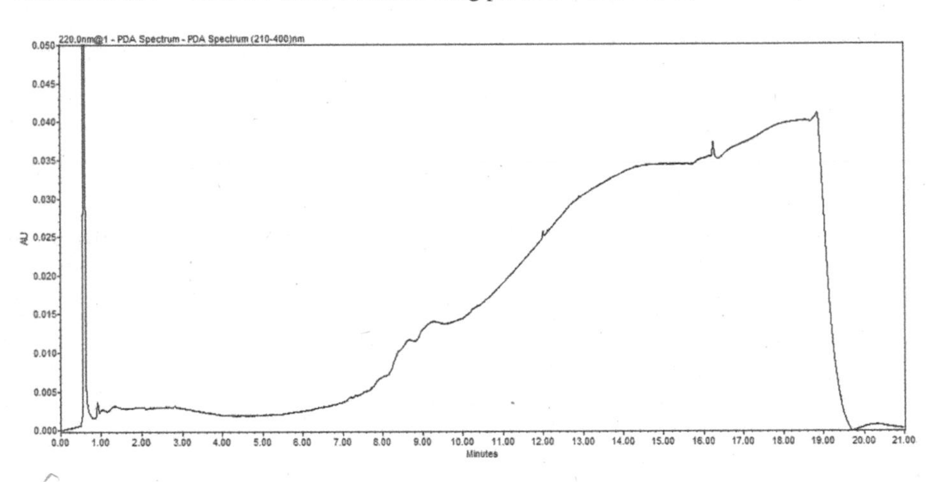

FIGURE 21.12 UPLC/UV/MS traces for drug product – contaminated unit ($\lambda = 220\,\text{nm}$)

TABLE 21.2
USP Signal to Noise for 0.05 μg/mL of Colorant in Methanol

Compound	RT (Min)	USP Signal to Noise		
Colorant	10.61	UV	APCI(+)	APCI(−)
		8	NA	NA

Chromatographers often must spend time readjusting analytical conditions when they install a new octa-decyl-silyl (ODS) column in their HPLC system, even if the new column replaces one from the same manufacturer. The column-to-column performance of ODS columns is very uniform. Analysts need to know how much time is required to readjust the analytical conditions. The carbon content (phase loading) of an ODS packing affects the concentration of the organic component needed in an aqueous mobile phase to elute a specific analyte. Two types of ODS columns require different mobile phases for the same analysis, primarily because they differ in the surface area of their silica base and to the extent the base is covered with the reversed phase. For example, a uniform brush monolayer of 3.4 μmoles/m^2 of dimethyl-octa-decyl-silane on LC-Si silica (170 m^2/g) and on silica (549–660 m^2/g) would produce almost a three-fold difference in the carbon content of the packings. The carbon content of differences in the column-to-column performance of a brand of ODS packing result from inconsistences in applying the phase to the silica base. The use of di- or trichloro-alkyl silanes will result in polymerization unless water is completely removed from the base silica and the reaction media. Such coverage cannot be exactly reproduced, which will produce batch-to-batch variability in packing performance. If the coverage of the silica surface is not complete, and a capping reaction is not included in the manufacturing process, the resulting HPLC packing material will have accessible surface silanol groups. Separations on a column filled with such a packing will occur through a mixture of normal phase and reversed-phase processes. In contrast, mono-chloro-di-methyl-alkyl silanes are used to manufacture LC-1, LC-8, LC-18, and LC-CN packings. This material does not polymerize, and it produces a uniform brush monolayer covering the silica base. Trimethylchlorosilane is used in a final capping step for packings. The product is a chemically uniform HPLC packing which is very reproducible from batch to batch.

Differences in the surface area and reversed-phase coverage of commercial ODS columns have two consequences. First, because most ODS packings exhibit batch-to-batch variability, the mobile phase used for analysis often must be adjusted when a new column is used. Second, the chromatographer who is transferring a separation from one ODS packing to another usually must adjust the mobile phase by trial and error. When an investigator using a column with mixed chromatographic properties changes to a true reversed-phase column (LC-18), it may be difficult to formulate a mobile phase that provides the same separations for all possible samples as the original column and mobile phase. The effort required to establish slightly different, but equal, true reversed-phase separations will be rewarded with far greater column-to-column reproducibility.

The capacity factor, k′, expresses the ratio of distribution of a compound between the stationary phase and the mobile phase in an HPLC column. The k′ for a compound will vary, depending on the characteristics of the stationary phase from which the compound is eluted, and on the concentration of the organic component in the mobile phase. If k′ is held constant when a compound is eluted from two or more columns, the mobile phases needed for constant analysis times can be compared. If several compounds with different characteristics are eluted at the same k' from two or more columns, then generalizations may be made about the mobile phases needed with each column. The k' value is obtained from the following formula: $k' = (t_r - t_0)/t_0$, where t_0 is the retention time of an unretained peak and t_r is the retention time of a retained peak.

Case Study: HPLC Columns Comparison

For practical purposes, compounds of interest usually are eluted within the k' range of 1–10. In the following comparison among manufacturers, standards were eluted from each packing with a series of organic solvents: water mixtures (i.e., 30%, 40%,

50%, and 60% organic solvent). The organic component: water mixture at which the k' equaled an arbitrarily chosen value of three was determined for each standard, with each packing, using three different organic solvents. Columns from Supeloc LC-18, Spherisorb ODS, Chromosorb® LC-7, Nucleosil® C18, LiChrosorb® RP-18, and Zorbax® BP-ODS were used from materials available commercially. These materials were assumed to be representative of the chosen packings. Several important characteristics of each packing were determined under standardized conditions and are presented in the table below. These data indicate that the eight ODS packings can be expected to differ in performance.

Benzene (C_6H_5H), phenol (C_6H_5OH), acetophenone ($C_6H_5COCH_3$), nitrobenzene ($C_6H_5NO_2$), methyl benzoate ($C_6H_5CO_2CH_3$), and toluene ($C_6H_5CH_3$) were eluted individually from each of the columns. These compounds were used as the analytical standards because each has a distinctive functional group and is easily detected by UV at 254 nm. Methanol: water, acetonitrile: water, and tetrahydrofuran: water mobile phases were selected because they are commonly used mixtures. Mobile phases were generated using a liquid chromatograph, and the reproducibility of k' values was verified. When methanol: water and acetonitrile: water mobile phases were used, the benzene derivatives eluted from the ODS packings in similar patterns. Although there were exceptions among a few of the compounds, progressively more methanol or acetonitrile in water generally was required to elute the analytes from Spherisorb, Chromosorb, μBondapak, SUPELCOSIL, LiChrosorb, Partisil, Nucleosil, and Zorbax BP-ODS, respectively. Some of the benzene derivatives were eluted from two or more of the packings with very similar mobile phases. In conclusion, the mean difference is approximately the extent of the mobile phase adjustment needed to adapt an analysis from one packing to another LC-18 packing (Table 21.3).

Column dimensions: Partisil 250×3.9 mm, μBondapak 300×4.6 mm, all others 150×4.6 mm. Column characteristics were determined under the following conditions: Mobile Phase: Methanol: water, 66:34 (v/v), Flow Rate: 1 mL/min.

Trademarks

μBondapak – Waters Associates

Chromosorb – Manville Products Corp.

LiChrosorb – EM Laboratories, Inc.

Nucleosil – Macherey-Nagel & Co.

Spherosil – Prolabo

SUPELCOSIL – Supelco, Inc.

Zorbax – E.I. du Pont de Nemours & Co., Inc.

The importance of determining component separations is related to three factors by a chromatographer comparing different ODS columns: the mobile phase composition needed

- elute analytes in reasonable time
- separation among analytes
- elution order of analytes

Increasing the percentage of the organic constituent in the mobile phase reduces analysis time, but it also reduces resolution. Therefore, rapid, high-quality separations

TABLE 21.3
Characteristics of Different Manufacturers' ODS Packings

Column	Carbon Content(%)	Theoretical Plates/ Meter (103)	Asymmetry at 10% of Peak Height	Back Pressure (psig)	Particle Diameter
SUPELCOSIL LC-18	10.76	73	1.03	950	5 μm (spherical)
Spherisorb ODS	7.33	71	1.33	1,290	5 μm (spherical)
Chromosorb LC-7	12.90	39	1.12	1,500	5 μm (irregular)
Nucleosil C18	15.28	80	1.29	1,800	5 μm (spherical)
LiChrosorb RP-18	20.13	54	1.18	1,830	5 μm (irregular)
Zorbax BP-ODS	13.86	63	1.16	850	7–8 μm (spherical)
Partisil 5 ODS	10	45	1.51	2,710	5 μm (irregular)
μBondapak C18	10	10	1.46	1,690	10 μm (irregular)

usually are best achieved with the column that requires the least organic constituent to elute a compound in a given period.

The separation of analytes is conveniently discussed in terms of the selectivity factor, between a specific reference analyte and each of the other components of the sample. The intersection between analytes should alert the analyst to the possibility that reversals in peak order can accompany changes in mobile phase composition. Reversals in the elution order of sample components can occur when the mobile phase composition is changed (especially if the packing has accessible surface silanol groups). As an example, selectivity values for a polar (methyl benzoate) and a nonpolar (benzene) compound are compared. Elution order of benzene and methyl benzoate occurs at 40–65% methanol with these ODS columns. Thus, separation of benzene and methyl benzoate on a Nucleosil column with a methanol: water mixture of 55:45 is attainable. For LC-18 column, 50:50 methanol: water mixture will do (Table 21.4).

HPLC COLUMNS

High Efficiency and Symmetrical Peaks

Three and five micrometer spherical silica particles with a narrow size distribution are acceptable bases for packings. This selection supports low back pressure and uniform flow through the columns. In addition, with this silica size selection, the necessary efficiency to separate complex samples and detect trace components (60,000–80,000 plates/ meter for 5 μm packings, 120,000+ plates/meter for 3 μm packings) is expected. For most column types, the asymmetry factor (measured at 10% peak height) is 0.85–1.20. All columns are supplied with fittings for connections to 1/16″ OD stainless steel tubing.

Reproducible analyses are affected by control of particle size, surface area (170 m^2/g for 5 μm, 100 Å silica), and pore diameter of silica, which play a major role in the quality of the stationary phase packing.

TABLE 21.4
Mobile Phase

Mobile Phase	C_6H_5OH		$C_6H_5COCH_3$		$C_6H_5NO_2$		C_6H_6		$C_6H_5CH_3$	
	Range	Average	Range	Average	Range	Average	Range	Average	Range	Average
(%) Methanol in Water										
40	0.15–0.18	0.17	0.38–0.50	0.43	0.48–0.60	0.54	0.66–0.90	0.76	1.49–2.17	1.74
50	0.19–0.23	0.22	0.45–0.51	0.49	0.57–0.71	0.63	0.74–1.10	0.94	1.45–2.40	1.98
60	0.23–0.28	0.26	0.50–0.64	0.55	0.67–0.79	0.73	0.97–1.28	1.10	1.71–2.56	2.07
70	0.26–0.31	0.29	0.50–0.60	0.56	0.70–0.85	0.76	1.02–1.33	1.13	1.67–2.40	1.93
% Acetonitrile in Water										
30	0.20–0.26	0.22	0.48–0.58	0.51	0.83–0.91	0.87	1.00–1.33	1.12	1.80–2.72	2.30
40	0.20–0.29	0.27	0.55–0.63	0.57	0.84–0.96	0.91	1.11–1.33	1.21	1.86–2.50	2.18
50	0.23–0.33	0.30	0.59–0.68	0.62	0.84–0.95	0.91	1.09–1.33	1.21	1.61–2.24	1.95
60	0.22–0.39	0.32	0.63–0.70	0.66	0.81–0.94	0.87	1.13–1.30	1.21	1.67–2.13	1.84
70	0.26–0.41	0.33	0.65–0.73	0.69	0.76–0.91	0.84	1.16–1.23	1.20	1.66–1.90	1.78
% Tetrahydrofuran in Water										
25	0.60–0.66	0.63	0.47–0.50	0.49	1.21–1.26	1.24	1.47–1.77	1.60	2.75–3.48	3.04
30	0.64–0.69	0.67	0.49–0.53	0.51	1.20–1.28	1.25	1.62–1.83	1.70	2.68–3.35	2.97
35	0.56–0.71	0.67	0.52–0.62	0.56	1.17–1.25	1.21	1.55–1.94	1.70	2.30–3.19	2.68
40	0.64–0.72	0.68	0.59–0.68	0.61	1.15–1.26	1.19	1.52–2.08	1.72	2.12–3.21	2.54
45	0.62–0.72	0.68	0.60–0.65	0.63	1.12–1.16	1.14	1.56–1.95	1.72	2.19–2.86	2.44
50	0.60–0.73	0.67	0.63–0.68	0.66	1.10–1.13	1.11	1.58–1.89	1.69	2.08–2.64	2.30

Universal Reversed-Phase HPLC Column Simplifies Method Development

Through treatment of the silica support and incorporating silanol shielding mechanism as part of the bonded phase chemistry, HPLC columns enable chromatographers to analyze strong acids or strong bases, using simple buffers and mobile phases. Unique chemical composition supports symmetric peaks, high efficiency, and special selectivity for a wide spectrum of analytes, which simplifies method development and makes it less time consuming. The packing materials of columns allow for gradient analyses and HPLC-MS. Chemically deactivated C18 columns separate small molecular weight compounds through the same reversed-phase mechanism that controls retention on traditional C18 / ODS reversed-phase columns.

Residual silanol groups on the surface of conventional reversed-phase silica packings often interact with acidic or basic compounds, resulting in low efficiency, tailing peaks, drifting retention time, and irreproducible separations. To obtain acceptable results, particularly for basic compounds, an amine modifier usually must be added to the mobile phase to minimize these interactions. Certain columns packings with specialty silica bonding can provide high efficiency and symmetric peaks for a wide range of strong acids and bases, with simple mobile phases. In preparing these columns, silica is thoroughly purified to produce a matrix reducing the number of highly active isolated free silanol groups. The accessibility of residual silanol groups is reduced by creating a very high surface coverage (> 5 μmoles/m^2). A unique silanol shielding layer, embedded near the silica surface, significantly reduces interactions between the analytes and remaining silanol groups. As a result, acids, bases, and neutral polar compounds, in various chemical classes, can be analyzed with high efficiency, symmetric peak shape, and constant retention time. Amine modifiers are unnecessary for chromatography of basic compounds. Method development is further simplified because buffers and organic modifiers have predictable effects on solute retention in the mobile phase.

Optimal resolution of acidic compounds containing a low ionic strength buffer added to acetonitrile or methanol is enough for analyzing even the strongest acids or bases. Efficiency and peak symmetry for sorbic and benzoic acids compare very well with values obtained with other deactivated reversed-phase columns. Because most pharmaceutical compounds are basic in nature, manufacturers of deactivated columns have focused on improving the chromatography of basic drugs.

Basic pharmaceuticals compounds with polar charged functional groups are attributed to the unique chemical structure of the bonded phase. The chromatograms can be different from the selectivity of conventional deactivated columns in which the order of elution changes. Analysts can take advantage of this unique chemical selectivity. Selective columns provide high efficiency for polar, nonpolar, and charged analytes, symmetric peaks for the most difficult compounds, hydrophobic selectivity, and special selectivity for polar and charged compounds targeting flat baseline for gradient analyses, preparative applications, and HPLC-MS.

Cardiac drugs containing acids, such as alkylbenzoic acids ($n = 0–5$), are retained much more strongly on C8 or C18 columns. Retention of the smallest chain acids is shorter on LC-18 columns than on LC-8 columns, but more hydrophobic acids are, as expected, retained longer on C18 columns. The slopes of the log k′ vs. n plots, which are proportional to the selectivity, for LC-18 are more selective for acids than LC-8 columns. The capacity factors, k′ and selectivity for alkyl-aniline compounds (aniline, 4-ethylaniline, 4-propylaniline, 4-butylaniline, 4-pentylaniline) are almost identical for traditional columns and deactivated LC-18 columns and are greater than values for LC-8 columns.

HPLC TROUBLESHOOTING AND GUIDE

Although HPLC method development has been improved by advances in column technology and equipment instrumentation, problems still arise. Systematic means of isolating, identifying, and correcting many typical problems require continued monitoring. Important segments of an HPLC system are essentially the same, whether it is a modular system or a more sophisticated unit. Problems affecting overall system performance can arise in each component.

Low sensitivity and rising baselines, noise, or spikes on the chromatogram can often be attributed to the mobile phase. Contaminants in the mobile phase are especially troublesome in gradient elution. The baseline may rise, and spurious peaks can appear as the level of the contaminated component increases. Water is the most common source of contamination in reversed-phase analyses. High purity distilled or deionized water should be used when formulating mobile phases. However, several common deionizers introduce organic contaminants into the water. To remove these contaminants, deionized water should be passed through activated charcoal or a preparative C18 column. HPLC grade solvents, salts, ion-pair reagents, and base and acid modifiers should be used. Trace contaminants can cause problems when using a high-sensitivity ultraviolet or fluorescence detector.

Because many aqueous buffers promote the growth of bacteria or algae, solutions should be freshly prepared and filtered (0.2 μm filter) before use. Filtering also will remove particles that could produce a noisy baseline or plug the column. Preventing microorganism growth is attained by adding about 100 ppm of sodium azide to aqueous buffers. Alternatively, these buffers may also be mixed with 20% or more of an organic solvent such as ethanol or acetonitrile.

To prevent bubbles in the system, degassing the mobile phase using an in-line degasser is a reasonable choice, but sparging with helium can be an alternative if the mobile phase does not contain any volatile components. Using ion-pair reagents should be carefully applied. The optimum chain length and concentration of the reagent must be determined for each analysis. Concentrations can be as low as 0.2 mM, or as high as 150 mM, or more. In general, increasing the concentration or chain length increases retention times. High concentrations (>50%) of acetonitrile or some other organic solvents can precipitate ion-pair reagents. Also, some salts of ion-pair reagents are insoluble in water and will precipitate. This can be avoided by using sodium-containing buffers in the presence of long-chain sulfonic acids (e.g., sodium dodecyl sulfate), instead of potassium-containing buffers.

Volatile basic and acidic modifiers, such as triethylamine (TEA) and trifluoracetic acid (TFA), are useful to recover a compound for further analysis. These modifiers use will also avoid problems associated with ion-pair reagents. They can be added to the buffer at concentrations of 0.1 to 1.0% TEA or 0.01 to 0.15% TFA.

Increasing the concentration may improve peak shape for certain compounds but can alter retention times.

Recycling the mobile phase used for isocratic separations (fixed composition and polarity) has become more popular in recent years as a means of reducing the cost of solvents, their disposal, and mobile-phase preparation time. An apparatus for solvent recovery uses a microprocessor-controlled switching valve to direct the solvent stream to waste when a peak is detected. When the baseline falls under the selected threshold, uncontaminated solvent is directed back to the solvent reservoir.

Isolating Pump Problems

HPLC pump must deliver a constant flow of solvent to the column over a wide range of conditions. Modern HPLC pumps incorporate single or dual piston, syringe, or diaphragm pump designs. Pumping system problems are usually easy to spot and correct. Some of the more common symptoms are erratic retention times, noisy baselines, or spikes in the chromatogram. Leaks at pump fittings or seals will result in poor chromatography. A sure sign of a leak is a buildup of salt deposit at a pump connection. Buffer salts should be flushed from the system daily with fresh deionized water. Pump seals require periodic monitoring or replacement as needed (Figure 21.13).

The injector rapidly introduces the sample into the system with the solvent flow. HPLC systems currently use variable loop, fixed loop, and syringe-type injectors. These are activated manually, pneumatically, or electrically. Mechanical problems involving the injector (e.g., leaks, plugged capillary tubing, worn seals) are easy to spot and correct. Pre-column filter is recommended to prevent plugging of the column. Variable peak heights, split peaks, and broad peaks can be caused by incompletely filled sample loops, incompatibility of the injection solvent with the mobile phase, or poor sample solubility. The injection solvent should of lower eluting strength than the mobile phase. Wash solution should be compatible with and weaker than the mobile phase. This is especially important when switching between reversed- and normal-phase analyses.

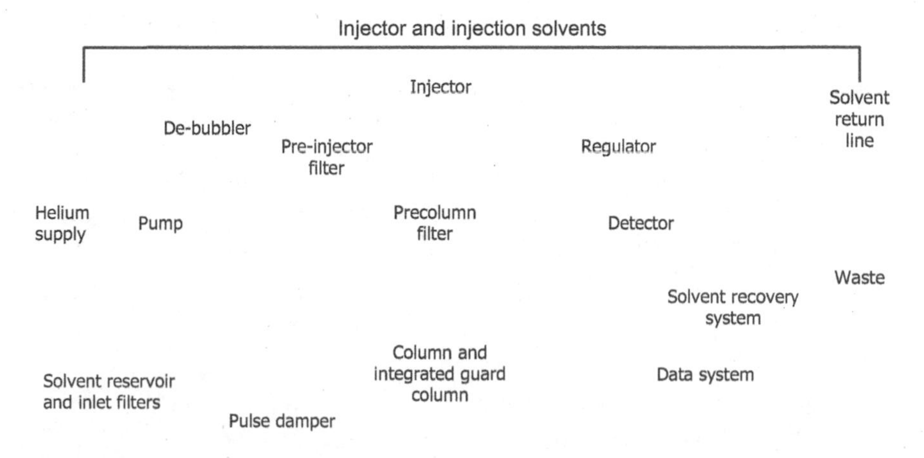

FIGURE 21.13 Components of an HPLC system.

Column Protection

Although not an integral part of most equipment, mobile-phase inlet filters, pre-injector and pre-column filters, and security guard columns greatly reduce problems associated with complex separations. It is recommended that all samples be filtered through 0.2 µm syringe filters. Filters and guard columns prevent particles and strongly retained compounds from accumulating on the analytical column. The useful life of these disposable pre-conditioning components depends on mobile phase composition, sample purity, pH, etc. As these devices become contaminated or plugged with particles, pressure increases, and peaks broaden or split.

Getting the Most from Analytical Column

Regardless of whether the column contains a bonded reversed or normal phase, ion exchange, affinity, hydrophobic interaction, size exclusion, or resin/silica-based packing, the most common problem associated with analytical columns is deterioration. Symptoms of deterioration are poor peak shape, split peaks, shoulders, loss of resolution, decreased retention times, and high back pressure. These symptoms indicate contaminants have accumulated on the column inlet, or there are voids, channels, or a depression in the packing bed.

Deterioration is more evident in higher efficiency columns. For example, a 3-micron packing retained by 0.5-micron frits (sintered porous metal frits are used in liquid chromatography columns to filter microbial and inorganic contaminants) is more susceptible to plugging than a 5- or 10-micron packing retained by 2 micron or larger frits. Proper column protection and sample preparation are essential to getting the most from each column. Overloading a column can cause poor peak shapes and other problems.

Solving Detector Problems

Detector problems fall into two categories – electrical and mechanical/optical. Electrical problems require service by the instrument manufacturer. Mechanical or optical problems usually can be traced to the flow cell. Detector-related problems include leaks, air bubbles, and cell contamination. These usually produce spikes or baseline noise on the chromatograms or low sensitivity. Some cells – especially those used in refractive index detectors – are sensitive to pressure. Flow rates or back pressures that exceed the manufacturer's recommendation will break the cell window. Old or defective lamps as well as incorrect detector rise time, gain, or attenuation will reduce sensitivity and peak height. Faulty or reversed cable connections can also be the source of problems.

Column Heater, Recorder

These components seldom cause problems with the system. If a fault is recognized, the operator should contact the supplier.

Keeping Accurate Records

Most problems develop gradually. Accurate record keeping, then, is vital to detecting and solving many problems. By keeping a written history of column efficiency, mobile phases used, lamp current, pump performance, etc., maintenance technicians can monitor system's performance and troubleshoot problems effectively. Records also help prevent mistakes, such as introducing water into a silica column or precipitating buffer in the system by adding too much organic solvent. Reliable records are the best way to ensure that a modification does not introduce problems. For problems relating to pumps, detectors, automatic samplers, and data systems, consult instrument manual's troubleshooting guide.

Further Recommendations

Referring to maintenance and troubleshooting sections of instrument manual is recommended. Modern HPLC systems often have self-diagnostic capabilities that help isolate a problem area within the instrument. For persistent problems relating to column analysis, operator should consult with manufacturer Technical Service Department.

Restoring Column's Performance

The following procedures should rejuvenate a column whose performance has deteriorated due to sample contamination.

Disconnect and reverse the column. Connect it to the pump, but not the detector. Follow the appropriate flushing procedure, using a flow rate that results in column back pressure of 1,500–4,500 psi, but never higher than the maximum recommended pressure in the manufacturer's instruction manual.

Nonbonded Silica Columns Exposed to Polar Solvent

Samples and mobile phases containing very strongly polar solvents, such as water or alcohols, can deactivate uncoated silica HPLC columns. This can drastically affect column performance, particularly solute retention, and selectivity. Even prolonged column flushing with a nonpolar solvent only partially restores column performance. A silica regeneration solution quickly and inexpensively restores silica column performance by removing trapped polar material. Pump the solution through the affected column for 10 minutes at a rate of 4 mL/minute, then flush with mobile phase for 10 minutes at a rate of 2 mL/minute.

Performance Evaluation Mixes for HPLC Columns

Well-defined test mixes enable operator to troubleshoot chromatographic problems, optimize system efficiency, and evaluate columns under conditions where their performance is understood.

Preventing and Solving Common Hardware Problems

Preventing Leaks

Leaks are a common problem in HPLC analyses. To minimize leaks in system, avoid interchanging hardware and fittings from different manufacturers. Incompatible fittings can be forced to fit initially, but the separation may show problems and repeated connections may eventually cause the fitting to leak. If interchanging is necessary, use appropriate adapters and check all connections for leaks before proceeding.

Highly concentrated salts (>0.2 M) and caustic mobile phases can reduce pump seal efficiency. The lifetime of injector rotor seals also depends on mobile phase conditions, particularly operation at high pH. In some cases, prolonged use of ion-pair reagents has a lubricating effect on pump pistons that may produce small leaks at the seal. Some seals do not perform well with certain solvents. Before using a pump under adverse conditions, read the instrument manufacturer's specifications. To replace seals, refer to the maintenance section of the pump manual.

Unclogging the Column Frit

A clogged column frit is another common HPLC problem. To minimize this problem from the start, use a pre-column filter and guard column. To clean the inlet, first disconnect and reverse the column. Connect it to the pump (but not to the detector), and pump solvent through at twice the standard flow rate. About 5–10 column volumes of solvent should be enough to dislodge small amounts of particulate material on the inlet frit. Evaluate the performance of the cleaned column using a standard test mixture.

Procedure to Open a Column

1. Disconnect the column from the system. To prevent the packing from oozing out of the column, perform subsequent steps as quickly as possible.
2. Using a vise and wrench, or two wrenches, carefully remove the inlet end fitting. If the frit remains in the fitting, dislodge it by tapping the fitting on a hard surface. If the frit stays on the column, slide it off rather than lift it off. This will help preserve the integrity of the packing bed.

   Modular columns may require a special tool to remove the frit cap.
3. Examine the old frit. Compression of the frit against the stainless-steel tubing will leave a ring around the edge on the column side of a properly seated frit. No ring can mean the ferrule is seated too near the tubing end. The resulting loose connection can leak silica or act as a mixing chamber.
4. Examine the packing bed. If it is depressed or fractured, operator needs a new column.
5. Replace the frit.
6. Replace the end fitting. Screw it down finger tight, then tighten 1/4 turn with a wrench.

Appendix I

API Terms

Active Pharmaceutical Ingredient (API) – Any substance or mixture of substances intended for use in the manufacture of a drug product and that, when used in the production of a drug, becomes an active ingredient of the drug product.

API Starting Material (API SM) – A raw material, intermediate, or an API that is used in the production of an API and that is incorporated as a significant fragment into the structure of the API. GMPs are applied from the introduction of the API starting material into the process.

Critical Process Parameter – A parameter whose value has a direct and measurable impact on the quality of the product.

Critical Quality Attribute – All attributes of the API that ensure that the drug product in which the API is used is safe and effective. These are ensured by the API specifications and GMP controls.

Drug Product – The dosage form in the final immediate packaging intended for marketing.

Excipients – Substances that are introduced in the manufacturing process and become part of the drug product.

Intermediate (I) – A material produced during the processing steps to an API, which must undergo further molecular change before it becomes an API. The final intermediate (FI) is one (1) molecular change away from the API.

"Key" Parameter – A parameter in pharmaceutical manufacturing, which has an influence on the properties of the material within the appropriate process step. "Key" parameters will be treated as critical parameters as specified.

Master Validation Plan (MVP) – A high-level document that establishes an umbrella validation plan for the entire process and is used to guide the project team in resource and planning.

Mother Liquor – The residual liquid that remains after the crystallization or isolation processes. It may be used for further processing.

Mother Liquor Recovery Cycle – The mother liquor recovery cycle is the manufacture of a series of batches where the first batch uses virgin materials, and the subsequent batches have the mother liquor from the previous batch added to them in order to recover the product from the mother liquor. The series of batches that start with the virgin material and end with the last recycle of the mother liquor is defined as a Mother Liquor Recovery Cycle.

Appendix II
Impurities – FDA Directive

DEPARTMENT OF HEALTH AND HUMAN SERVICES

Food and Drug Administration
Silver Spring, MD 20993

General Advice

Dear Sir or Madam:

This letter is to inform applicants with an approved or pending application for an angiotensin II receptor blockers (ARB) drug product (DP), as well as holders of related drug master files (DMFs), of FDA concerns related to the presence of one or more toxic impurities in some ARB drugs. This general advice letter summarizes FDA findings to date and provides recommended actions to take to ensure that your drug product, drug substance/active pharmaceutical ingredient (API), and raw materials are absent of these impurities or below our recommended limit.

Background

In June 2018, FDA was informed of the presence of an impurity, identified as N-nitrosodimethylamine (NDMA), from one valsartan API producer. Since then, FDA has determined that other types of nitrosamine compounds, e.g., N-nitrosodiethylamine (NDEA), are present at unacceptable levels in APIs from multiple API producers of valsartan and other drugs in the ARB class. FDA has and will continue to provide periodic updates on this problem on its website (https://www.fda.gov/drugs/drugsafety/ucm613916.htm). FDA continues to collaborate with other drug regulatory agencies around the world to address this problem, and FDA is committed to working with manufacturers and applicants impacted by this problem to bring this to a full and timely resolution. FDA and other regulators, as well as many manufacturers, have developed and are using methods validated to detect and quantify a variety of nitrosamine impurities. FDA has posted these methods on its website. FDA has also issued Information Request letters to application and DMF holders to request specific information and samples of drugs thought to be at risk for the presence of a nitrosamine impurity.Nitrosamine compounds are potent genotoxic carcinogens in several nonclinical species and are classified as probable human carcinogens by the International Agency for Research on Cancer (IARC). In fact, "N-nitroso" compounds are identified as a "cohort of concern" in internationally harmonized guidance, ICH M7, *Assessment and Control of DNA Reactive (Mutagenic) Impurities in Pharmaceuticals to Limit Potential Carcinogenic Risk.* ICH M7 recommends that known mutagenic carcinogens, such as nitrosamines, be controlled at or below the acceptable cancer risk level. Due to their known potent carcinogenic effects, and because it is feasible to limit these impurities by taking reasonable steps to prevent or eliminate their presence, FDA has determined that there is no acceptable specification for nitrosamines in ARB API and DP. Therefore, FDA advises that nitrosamines should be absent (i.e., not detectable as described below) from ARB API and ARB drug products. As an initial measure, FDA published "interim acceptable limits" for these nitrosamine impurities in ARBs. ARB DS or DP with levels of impurities exceeding these interim limits were recommended for recall from the market. FDA has used the interim limits to guide immediate decision-making for product recalls to balance the risks of potential long-term carcinogenic risk and disruption to clinical management of patients' hypertension and heart failure. FDA is now seeking the information outlined below to ensure that ARB API and DP entering the marketplace have no detectable nitrosamines.Recent information gathered by FDA suggests several general causes of the presence of a nitrosamine impurity in ARB APIs. First, we now know that nitrosamine impurities can form during API processing under certain processing conditions and in the presence of some types of raw materials and starting materials. These materials include intermediates that are not purged in subsequent steps of the API process. A second cause appears to be from the use of contaminated raw materials used in the manufacturing process. Recovered materials, such as recovered solvents and catalysts, may pose a risk for nitrosamine formation due to the presence of amines in the solvents or catalysts sent for recovery and the subsequent quenching of these materials with nitrous acid to destroy residual azide without adequate removal. Independent recovery facilities may comingle solvents/catalysts from various customers or not perform adequate cleaning of equipment between customers. A similar cause may be from contaminated starting materials, including intermediates supplied by a vendor, that use processing methods or raw materials causing the formation of nitrosamines in their material. Contamination from vendor-sourced raw materials and starting materials/intermediates is particularly challenging because an API producer whose process is not capable of forming a nitrosamine compound may not realize their process is at risk to the presence of such impurities. FDA is aware that some ARB producers have identified a nitrosamine in their finished API, even though they are using processes incapable of forming a nitrosamine impurity.The multiple causes listed above can occur in the same API process. The typical tests for API purity, identity, and known impurities are unlikely to detect the presence of a nitrosamine impurity. Further, each failure mode could result in different nitrosamines, different amounts, or undetectable amounts of nitrosamine impurities in different batches from the same process and API producer.

Accordingly, FDA advises:

1. ARB DP manufacturers test representative samples of each drug product batch, or alternatively a representative sample of each API lot used in each drug product batch, they have produced for the US market to determine whether any contain a detectable amount (defined below) of a nitrosamine impurity. Testing should include DP batches already distributed that have not yet reached their labeled expiration date as well as those not yet distributed by the DP manufacturer. Any DP batch already in distribution, as of the date of this letter, with a nitrosamine level that exceeds the FDA published interim acceptable limit should be recalled, if distributed, or quarantined pending appropriate disposition if not distributed. Any DP product batch found to contain a detectable nitrosamine impurity that is below the interim acceptable limit should not be released by the DP manufacturer for distribution unless FDA agrees that distribution is warranted to prevent or mitigate a shortage of a medically necessary drug (or if for export from the US, a shortage determination was made by the importing country's national regulatory agency).
2. ARB DP manufacturers test representative samples of each API batch in their possession to demonstrate the absence of nitrosamines prior to use in DP manufacturing. In addition, DP manufacturers should test each API lot received from each supplier before releasing the API for use until the DP manufacturer has verified that the supplier can consistently produce API without a detectable nitrosamine (as defined below) in accordance with the CGMP regulations at 21 CFR 211 subpart E; see also FDA guidance for industry, ICH Q10, *Pharmaceutical Quality System.*
3. ARB applicants (or DP manufacturers on their behalf) report to FDA a finding of detectable nitrosamine in either an API lot or in a DP batch whether distributed. The Field Alert Report (FAR) regulation for ANDAs and NDAs (21 CFR 314.81) requires such reports for distributed batches. If not distributed, reporting a finding of detectable nitrosamine to the FAR system will assist FDA in timely remediation. For instructions on submitting a FAR, please refer to the draft guidance document *Field Alert Report Submission: Questions and Answers Guidance for Industry* at https://www.fda.gov/downloads/Drugs/GuidanceComplianceRegulatoryInformation/Guidances/UCM613753.pdf.
4. Applicants report to FDA a summary of the testing performed, as requested above, for the presence of any nitrosamine impurities in batches distributed in the US or exported from the US that are within their labeled expiration, even if recalled. We request a table be submitted to each application, if not already provided in previous correspondence to the application, with the following information for each batch number that is sampled and tested:
   - product name (identify whether API or DP batch)
   - labeled strength (if DP batch)
   - date of manufacture
   - labeled expiration date
   - name of test method
   - amount and type of nitrosamine detected, if any, or "none detected."

Data should be submitted to the application as a "General Correspondence"; the words "nitrosamine-related" should be prominently displayed on the cover letter.

5. Applicants of pending ARB applications should provide a written statement as General Correspondence declaring that the API supplier provides API that does not contain any detectable nitrosamine impurities. No further action in response to this letter is needed if this information has already been submitted to the application or referenced DMF.
6. ARB API producers test representative samples of each batch of an ARB API to determine whether any contain detectable nitrosamine impurities. Testing should include API batches distributed and within expiry, labeled with a "retest by" date, and those not yet distributed. Any API batch containing a nitrosamine impurity above the interim acceptable limits should be recalled, if distributed, or dispositioned as not suitable for use in DP intended for the US market. If detected below the interim acceptable limit, the batch should not be distributed for use in DP intended for the US market unless FDA agrees that such use is warranted to prevent or mitigate a US shortage of a medically necessary drug.
7. API batches may be reprocessed, reworked, and/or reconditioned to be rendered absent of a detectable nitrosamine impurity as provided for in existing policies for amending or supplementing and controlling such operations. If a batch is found to contain nitrosamines and is reprocessed or reworked in any way, this should be reported to the DMF and/or application. Please note that such amendments may have user fee goal date implications and will be assessed accordingly.
8. ARB API producers evaluate each process for the potential to form a nitrosamine impurity and if at risk, make changes necessary to prevent nitrosamine formation. If the process cannot be changed to prevent nitrosamine formation, FDA will permit the use of a robust purging/elimination step(s) if it includes an appropriately sensitive test to verify that the resulting intermediate or API does not contain a detectable nitrosamine impurity. See existing FDA guidance in ICH Q7 *Good Manufacturing Practice Guidance for Active Pharmaceutical Ingredients*, ICH Q11 *Development and Manufacture of Drug Substances*, and ICH M7.
9. Batch testing to verify no detectable nitrosamine in the API should continue unless the API producer has demonstrated their process is not at risk for producing detectable nitrosamine in accordance with guidance (see, e.g., ICH Q7). This includes demonstrating that:
   - starting materials, including vendor-supplied intermediates, have no detectable nitrosamines or such amounts can be purged such that the API contains no detectable amounts of nitrosamines, and
   - raw materials used in the process, including recovered solvents and catalysts, contain no detectable amounts nitrosamines.
10. ARB API producers should voluntarily report the finding of a nitrosamine impurity to FDA in a Field Alert Report even if the contaminated material was not used in API processing. FDA will review the reports to determine if

other API producers are unknowingly at risk to nitrosamine contamination and notify accordingly.

11. ARB API producers report to FDA a summary of the testing performed, as requested above, for the presence of any nitrosamine impurities in batches distributed in the US, whether directly as an API or after incorporation into a DP for the US, that are within their labeled expiration or retest-by date, even if recalled. We request a table be submitted to each DMF, if not already provided in previous correspondence to the DMF, with the following information for each batch number that is sampled and tested:
    - API name
    - date of manufacture
    - labeled expiration or retest-by date
    - name of test method
    - amount and type of nitrosamine detected, if any, or "none detected."

    Data should be submitted to the Drug Master File as "Quality/Controls"; the words "nitrosamine-related" should be prominently displayed on the cover letter.
12. ARB API producers report to each ARB DMF information about each independent facility that recovers materials used in ARB production for the past two years. We request the following information about such facilities: business name, address, name of recovered material, and the month and year the recovery facility has been performing recovery operations for the ARB.

FDA will, to the extent possible, expedite review of amendments and supplements for manufacturing changes required to eliminate or limit a nitrosamine impurity, or when needed to prevent or mitigate a drug shortage.For the testing requested in this letter, the detectable amount of nitrosamine impurity should be based on one of the following:

1. The limit of detection established in one of the FDA's published methods.
2. A method published by another regulatory agency that is equivalent to FDA's method(s).
3. Any appropriately developed and validated method capable of a LOD and Limit of Quantitation (LOQ) equivalent to a method published by FDA.

The interim acceptable limits and FDA published methods, as well as other FDA information on this issue, are available at https://www.fda.gov/Drugs/DrugSafety/ucm613916.htm.FDA published methods have been validated to detect and quantify NDMA and NDEA in all ARB APIs and some ARB DP formulations. FDA may update existing methods or post new methods once validated for use in detecting other nitrosamines in DPs and APIs. FDA may also update its published methods to improve their limits of detection and/or quantitation; if updated, FDA expects that manufacturers will update their methods to achieve comparable limits and apply the new LOD, if any, in making decisions about batch suitability.You should share this letter with your suppliers (e.g., solvent recovery vendors and starting material suppliers) and contract manufacturers.

References

Lieberman H, Loachman L, Schwartz JB, eds. *Pharmaceutical Dosage Forms: Tablets*, Volumes 1, 2 & 3, 2nd ed; Marcel Dekker, Inc., New York, 1990.

Seminar handouts from *The Center for Professional Advancement.* course on Modern Granulation, Tableting & Capsule Technology; August 1–4, 1988.

Seminar handouts from *Fluid Bed Training Seminar.* David M. Jones of Glatt Air Techniques, Inc; December 8–9, 1996.

Seminar handouts from *Film-Coating Technology.* Louis R. Palermo of Colorcon; February 1997.

Controlled Release Systems: Fabrication Technology Vol. I, Dean S.T. Hsieh, CRC Press; 1988, rev 2, 8/30/97.

Bibliography

1. *Cleaning Validation: Practical Compliance Solutions for Pharmaceutical Manufacturing*, Volume 1 by Destin LeBlanc.
2. *Cleaning Validation: Practical Compliance Solutions for Pharmaceutical Manufacturing*, Volume 2, by Destin LeBlanc.
3. *Cleaning Validation: Practical Compliance Solutions for Pharmaceutical Manufacturing*, Volume 3, by Destin LeBlanc.
4. *Cleaning Validation: Practical Compliance Solutions for Pharmaceutical Manufacturing*, Volume 4, by Destin LeBlanc.
5. *Cleaning and Cleaning Validation*, Volumes 1 and 2, Edited by Paul Pluta.
6. *Cleaning and Cleaning Validation: A Biotechnology Perspective.*
7. *Cleanroom Micro*biology, by Tim Sandle, R. Vijaya Kumar.
8. *Cold Chain Chronicles: A practitioners outside-the-box perspectives on the importance of temperature-sensitive drug stewardship*, by Kevin O'Donnell.
9. *Combination Products: Implementation of cGMP Requirements*, Edited by Lisa Hornback.
10. *Computerized Systems in the Modern Laboratory: A Practical Guide*, by Joseph Liscouski.
11. *Confronting Variability: A Framework for Risk Assessment*, Edited by Diane Petitti, Richard Prince.
12. Container/Closure Integrity Assessment A Compilation of Papers from the PDA Journal of Pharmaceutical Science and Technology.
13. *Contamination Control in Healthcare Product Manufacturing*, Volume 1 Edited by Russell Madsen, Jeanne Moldenhauer.
14. *Contamination Control in Healthcare Product Manufacturing*, Volume 2, Edited by Russell Madsen, Jeanne Moldenhauer.
15. *Contamination Control in Healthcare Product Manufacturing*, Volume 3, Edited by Russell Madsen, Jeanne Moldenhauer
16. *Contamination Control in Healthcare Product Manufacturing*, Volume 4, Edited by Russell Madsen, Jeanne Moldenhauer.
17. *Contamination Control in Healthcare Product Manufacturing*, Volume 5, Edited by Russell Madsen, Jeanne Moldenhauer.
18. *Contamination Prevention for Nonsterile Pharmaceutical Manufacturing*, By Andrew Dick.
19. *Effective Implementation of Audit Programs*, By Miguel Montalvo.
20. *Encyclopedia of Rapid Microbiol Methods*, Volume 4, Edited by Michael Miller.
21. *Encyclopedia of Rapid Microbiological Methods*, Volumes 1, 2 and 3, Edited by Michael Miller.
22. *Environmental Monitoring*, Edited by Jeanne Moldenhauer.
23. *Environmental Monitoring: A Comprehensive Handbook*, Volume 1, 2 and 3, Edited by Jeanne Moldenhauer.
24. *Environmental Monitoring: A Comprehensive Handbook*, Volume 4, Edited by Jeanne Moldenhauer.
25. *Environmental Monitoring: A Comprehensive Handbook*, Volume 5, Edited by Jeanne Moldenhauer.
26. *Environmental Monitoring: A Comprehensive Handbook*, Volume 6, Edited by Jeanne Moldenhauer.

27. *Environmental Monitoring: A Comprehensive Handbook*, Volume 7, Edited by Jeanne Moldenhauer.
28. *Environmental Monitoring: A Comprehensive Handbook*, Volume 8, Edited by Jeanne Moldenhauer.
29. *Environmental Monitoring: A Comprehensive Handbook*, Volumes 4, 5, 6 and 7 Edited by Jeanne Moldenhauer.
30. *Ethylene Oxide Sterilization Validation and Routine Operations Handbook* by Anne Booth.
31. *FDA Warning Letters: Analysis and Guidance*, By Jeanne Moldenhauer.
32. *Fungi: A Handbook for Life Science Manufacturers and Researchers*, Edited by Jeanne Moldenhauer.
33. *GMP in Practice: Regulatory Expectations for the Pharmaceutical Industry*, Fifth Edition, by Tim Sandle, James Vesper.
34. *Global Sterile Manufacturing Regulatory Guidance Comparison.*
35. *Good Distribution Practice: A Handbook for Healthcare Manufacturers and Suppliers*, Volume 1, Edited by Siegfried Schmitt.
36. *Good Distribution Practice: A Handbook for Healthcare Manufacturers and Suppliers*, Volume 2, Edited by Siegfried Schmitt.
37. *Hosting a Compliance Inspection* by Janet Gough.
38. *Introduction to Environmental Monitoring in Pharmaceutical Areas* by Michael Jahnke.
39. *Laboratory Design: Establishing the Facility and Management Structure* Edited by Scott Sutton.
40. *Lessons of Failure: When Things Go Wrong in Pharmaceutical Manufacturing* Edited by Maik Jornitz, Russell Madsen.
41. *Lifecycle Risk Management for Healthcare Products: From Research Through Disposal*, Edited by Edwin Bills, Stan Mastrangelo.
42. M*edia Fill Validation Environmental Monitoring During Aseptic Processing* by Michael Jahnke.
43. *Method Development and Validation for the Pharmaceutical Microbiologist*, By Crystal Booth.
44. *Microbial Control and Identification: Strategies Methods Applications*, Edited by Mary Griffin, Dona Reber.
45. *Microbial Identification: The Keys to a Successful Program*, Edited by Mary Griffin, Dona Reber.
46. *Microbial Risk Assessment in Pharmaceutical Clean Rooms* (single user digital version), By Bengt Ljungqvist, Berit Reinmueller.
47. *Microbial Risk and Investigations*, Edited by Karen McCullough, Jeanne Moldenhauer.
48. *Microbiological Culture Media: A Complete Guide for Pharmaceutical and Healthcare Manufacturers*, By Tim Sandle.
49. *Microbiological Monitoring of Pharmaceutical Process Water* by Michael Jahnke.
50. *Microbiology in Pharmaceutical Manufacturing*, Second Edition, Volumes 1 and 2, Edited by Richard Prince.
51. *Microbiology, and Engineering of Sterilization Processes*, Twelfth Edition 2007, By Irving Pflug.
52. PDA Technical Series: Endotoxin Analysis and Risk Management.
53. PDA Technical Series: Pharmaceutical Glass PDF Single user.
54. *Pharmaceutical Contamination Control: Practical Strategies for Compliance*, Edited by Nigel Halls,
55. *Pharmaceutical Legislation of the European Union, Japan, and the United States of America - An Overview*, By Denyse Baker, Joanne Hawana, Takayoshi Matsumura, Edited by Barbara Jentges.

56. *Pharmaceutical Outsourcing: Quality Management and Project Delivery* Edited by Trevor Deeks, Karen Ginsbury, Susan Schniepp.
57. *Pharmaceutical Quality* Edited by Richard Prince.
58. *Pharmaceutical Quality Control Microbiology: A Guidebook to the Basics* by Scott Sutton.
59. *Phase Appropriate GMP for Biological Processes: Pre-clinical to Commercial Production* Edited by Trevor Deeks.
60. *Practical Aseptic Processing Fill and Finish*, Volumes 1 and 2 Edited by Jack Lysfjord.
61. *Quality by Design: Putting Theory Into Practice* Edited by Siegfried Schmitt.
62. *Radiation Sterilization: Validation and Routine Operations Handbook* by Anne Booth.
63. *Rapid Sterility Testing* Edited by Jeanne Moldenhauer.
64. *Recalls of Pharmaceutical Products: Eliminating Contamination and Adulteration Causes* by Tim Sandle.
65. *Recent Warning Letters Review for Preparation of a Non-Sterile Processing Inspection*, Volume 2 By Jeanne Moldenhauer.
66. *Recent Warning Letters Review for Preparation of an Aseptic Processing Inspection*, Volume 1 By Jeanne Moldenhauer.
67. *Risk Assessment and Management for Healthcare Manufacturing: Practical Tips and Case Studies* by Tim Sandle.
68. *Risk Assessment and Risk Management in the Pharmaceutical Industry: Clear and Simple* by James Vesper.
69. *Risk-Based Compliance Handbook* by Siegfried Schmitt.
70. *Risk-Based Software Validation* by Janet Gough, David Nettleton.
71. *Root Cause Investigations for CAPA: Clear and Simple* by James Vesper.
72. *SOPs Clear and Simple: For Healthcare Manufacturers* by Brian Matye, Jeanne Moldenhauer, Susan Schniepp.
73. *Software as a Service (SaaS): Risk-Based Validation with Time-Saving Templates* by Janet Gough, David Nettleton.
74. *Square Root of (N) Sampling Plans: Procedures and Tables for Inspection of Quality Attributes* by Joyce Torbeck, Lynn Torbeck.
75. *Steam Sterilization: A Practitioner's Guide* Edited by Jeanne Moldenhauer.
76. *Sterility Testing of Pharmaceutical Products* by Tim Sandle.
77. *Systems Based Inspection for Pharmaceutical Manufacturers* Edited by Jeanne Moldenhauer.
78. *Technology and Knowledge Transfer: Keys to Successful Implementation and Management* Edited by Mark Gibson, Siegfried Schmitt.
79. *The Bacterial Endotoxins Test: A Practical Guide* by Karen McCullough.
80. *The External Quality Audit* by Janet Gough, Monica Grimaldi.
81. *The Internal Quality Audit* by Janet Gough, Monica Grimaldi.
82. *Thermal Validation in Moist Heat Sterilization* Edited by Jeanne Moldenhauer.
83. *Torbeck's Statistical Cookbook for Scientists and Engineers* by Lynn Torbeck.
84. *Trend and Out-of-Trend Analysis for Pharmaceutical Quality and Manufacturing Using Minitab* by Lynn Torbeck.
85. *Validating Enterprise Systems: A Practical Guide* by David Stokes.
86. *Validation Master Plan: The Streetwise Downtown Guide* by Trevor Deeks.
87. *Validation by Design: The Statistical Handbook for Pharmaceutical Process Validation* by Lynn Torbeck.
88. *Validation of Analytical Methods for Biopharmaceuticals: A Guide to Risk-Based Validation and Implementation Strategies* by Stephan Krause.
89. *Visual Inspection and Particulate Control* by Scott Aldrich, Roy Cherris, John Shabushnig
90. *Why Life Science Manufacturers Do What They Do in Development, Formulation, Production and Quality: A History* by Lynn Torbeck

Index